For a number of years Michael Tan's writings about contemporary arguments and developments in science and technology education have, through his breadth and depth of thinking across a very wide range of relevant fields, challenged and informed me in very helpful ways. Rather than me write about why this has been so, I suggest you turn to the final paragraph of Chapter 1 for a hint of the spectrum of fundamental ideas Michael will present to the reader as they work through this important volume.

—*Richard Gunstone*, Emeritus Professor of Science and Technology Education, Monash University

Michael Tan challenges us to rethink science and STEM education – indeed education more generally – to address the needs of a new century. Focusing on moving beyond schooling as reproduction to advocate innovation, creativity and forms of knowledge that serve the new millennium, in this book Michael has offered productive lines of attack for a reenvisaged curriculum. Pulling together a wide array of scholars to challenge our thinking and seek ways forward, he has crafted a narrative that is both educative and stimulating.

—*Russell Tytler*, Alfred Deakin Professor and Chair of Science Education, Deakin University

MAKERSPACES, INNOVATION AND SCIENCE EDUCATION

This book provides an overview to a range of theories in science and technology that inform the different ways in which makerspaces can be educative. Makerspaces are an indispensable site for science, technology, engineering, and mathematics (STEM) instruction and pose novel risks and opportunities for STEM instruction. Educators are likely to reach towards activities that have a high degree of engagement, but this might result in observations like 'it looks like fun, but *what* are they learning?'.

Beginning from the question of how we know what we know in science, the author asserts that understanding scientific knowledge requires us to know more than the abstract concepts typically presented in schools. The social and material aspects of knowledge are also important—these take the form of questions such as: What is the interplay between knowledge and power? How do we understand that we can have a 'feel' for materials and artefacts that we cannot completely describe in words? How do we know what ideas ought to be made real through technology and engineering? Significantly, this book also discusses the ethical dimensions of STEM education, in thinking about the kinds of STEM education that could be useful for open futures.

This book will be useful to graduate students and educators seeking an expansive view of STEM education. More generally, these ideas outline a possible new strategy for a vision of school that is not merely training or preparing students for work. Education needs to also prepare students for sociopolitical participation, and with STEM being central to our contemporary lives, this book provides insights for how this can happen in makerspaces.

Michael Tan is a Lecturer (Research Scientist) at the National Institute of Education, Nanyang Technological University, Singapore. He researches the nature of scientific and technological knowledge, and the social consequences of the differential distribution of this knowledge through schools.

MAKERSPACES, INNOVATION AND SCIENCE EDUCATION

How, Why, and What For?

Michael Tan

Routledge
Taylor & Francis Group

LONDON AND NEW YORK

Cover image: Getty Images

First published 2022
by Routledge
4 Park Square, Milton Park, Abingdon, Oxon OX14 4RN

and by Routledge
605 Third Avenue, New York, NY 10158

Routledge is an imprint of the Taylor & Francis Group, an informa business

British Library Cataloguing-in-Publication Data
A catalogue record for this book is available from the British Library

Library of Congress Cataloging-in-Publication Data
A catalog record has been requested for this book

ISBN: 9780815361534 (hbk)
ISBN: 9780815361541 (pbk)
ISBN: 9781351116220 (ebk)

DOI: 10.4324/9781351116220

Typeset in Bembo
by KnowledgeWorks Global Ltd.

CONTENTS

1

INTRODUCTION

Makerspaces have become a current fashion in educational circles in the last decade. Along with its rise has been the increasing popularity of the concept of 'STEM': science, technology, engineering, and mathematics. Rather understandably, a lot of the attention in research has been on the diverse ways in which educators can use the huge range of new technologies and techniques to engage young learners with apparently new learning goals. It has not helped that the memory of the turn of a new century has been rather recent and that there has seemed to be a large degree of disruptive change in these past two decades. These circumstances conspire to give us all the collective illusion that many of the old problems of the previous century (and earlier) can be safely ignored if only because the problems that we now face are altogether seemingly more urgent. With almost non-stop breathless coverage of technologies such as 'artificial intelligence', 'the internet of things', and 'the fourth industrial revolution', it can be very easy to sell a largely ignorant public the narrative that STEM and makerspaces constitute the future of science education. Science educators can therefore continue to keep their positions of expertise and authority—'trust us to know how to teach your children to remain relevant in these rapidly changing times'.

This book does *not* intend to do these things. Even as I announce in the title that I will be writing about the 'how' of makerspaces and science education, I do not intend to prescribe specific tactics that have been 'empirically proven' to work in the classroom. I imagine this book to be useful for fellow science educators who sense that there must be more to science education than just communicating the facts and concepts of science that we have accumulated over the years. In ambition, I stand in alliance with people who share the enlightenment ideal of a kind of science that can assert and support a particular kind of truth against a dominant order which is ultimately judged unworthy. Education, for me and many others, is a deeply political act and involves the liberation of individuals

DOI: 10.4324/9781351116220-1

and communities: we become liberated when we are no longer ruled by superstition, meaningless rituals, and irrational beliefs. We are liberated when we no longer *need* to rely on the authority of specific trusted institutions whose warrants are based on tradition or unquestionable faith. We are liberated when we can imagine a better way to organise life and can take steps that can make these alternatives a reality. This book is an attempt to think about the *why* and *what for* of science education, especially in makerspaces, conceptualised broadly as any space that facilitates the open-ended exploration, tinkering, and construction of artefacts for purposes largely related to the goals of understanding STEM.

Yes, we do need the conceptual knowledge and facts of science, but that is not enough. The scientific way of thinking probably gives us our best chance today of getting at truths and preventing us from fooling ourselves, but I worry that excessive specialisation and reductionism ubiquitous in scientific practice encourages science *education* research into a similar reductionist mindset. In doing so, we may think that what is important are the details and lose the forest for the trees.

★ ★ ★

We live in strange times. This ought not to be the opening sentence in an academic book purporting to survey the landscape on science education and to propose possible ways forward. Yet, here we are. But then, perhaps it is completely polemical to claim so; perhaps there never has been any such time where life was not strange to its inhabitants. Nonetheless, the quality of this specific strangeness is quite different from other times. We have a former reality-TV personality occupying an important government office in a powerful country, bullying individuals through a social media platform, and threatening entire nations with weapons of mass destruction. We have situations where 'fair and balanced' reporting means that one person's ignorance is equivalent to another's expertise. 'Post-truth' recently became word of the year of a particular famous lexicographic service; through it all, the incessant drumbeat of capitalism demands that we all dance the same, boring dance, justifying all sorts of decisions based on how much 'wealth' we can 'create'.

Science has been largely complicit in this state of affairs. Schools have also been part of this problem. This book is an extended essay investigating why this is so, and what could be done to get us out of this mess. To be sure, we probably attribute too much to what schools can do to change societies. However, it is certainly no coincidence that universities have always been the hotbeds of revolution. In contemporary societies, schools and universities are probably the only places left where, at least in intent, we regularly ask people to imagine what else society could be, before we crush their dreams in the industrialised shredder of the standardised assessment complex. Between the ennui of more-of-the-same solutions looking for problems and schools stuck in loops completely unable to move past their own founding narratives, we urgently need new ways of thinking about what science can do, what science should do, and how we ought to educate if we care enough about real solutions to the problems we currently face.

We need to become clearer about the nature of the problems that we face, and how it is that we can get schools to attend to these problems, or not. Chief among these must be the degree of complication and complexity that confronts anybody who tries to investigate how the world can be a better place. While there probably never has been straightforward answers, the problems are likely to be more complicated now than ever before due to the amplifying effects of technological devices. Well-intentioned 'innovations' produce undesired side effects that are patched with further 'innovations', which produce further undesired side effects, until 'innovators' realise that we should just give up and celebrate the 'disruptiveness' as a main, not side effect of these technological patches. All through this, we tacitly consent to have technologists and scientists drive the envisioning of the state of the future—bright shiny futures of flying cars, ubiquitous communication, and impossible architecture, not to mention the ever-optimistic sci-fi trope of a post-scarcity society, and the boring work done by machines. In the words of the screenwriters for one such movie, we really ought to consider if we are producing 'scientists [...] so preoccupied with whether or not they could, they didn't stop to think if they should'.

It is not hard to see how much of the world we live in requires some degree of scientific understanding if we are not to live in a world of magic.[1] Conversely, for large segments of the population who do not understand how the devices around them work, one supposes that the frustration and disempowerment that occurs when devices do not work as they have can be serious indeed. This is, only in part, an argument for a kind of science education that connects representational competencies with its performative counterpart. This book is concerned with developing the educative value proposition not only for makerspaces but also for making, more generally speaking, as a learning activity and a context that advances a particular vision of science and its place in society.

This vision, as sketched out above, has us consider science, not as a standout in its hubristic excess or as a variation of how Ernest Rutherford would have it: 'In science there is only physics, all else is stamp collecting'. This is not to lower science's place in relation to other forms of knowledge, but rather to acknowledge the role of these differential contributions to our holistic understanding. I am certainly not alone in considering this perspective—researchers have drawn implications for schooling emanating from studies in the sociology of scientific knowledge (SSK) or science and technology studies (STS). The main concern, which I believe still has yet to be conclusively resolved, is the status of scientific knowledge: what makes scientific knowledge special, and what are we to make of it? A useful heuristic here is the characterisation by Collins and Evans [cite] of the three waves of STS, from positivistic excess over to postmodern disenchantment, and now a third wave which acknowledges the social constructedness of scientific knowledge and all its faults, but which nonetheless argues for the importance of upholding scientific values. This third wave, which is an analogous movement to social realist thought in epistemology, drives much of the argument in the first part of this book. It is my contention here that studies in science education have not fully developed the implications emanating from this more current interpretation of the ontology and epistemology of science and scientific knowledge.

Even if we do take ontological and epistemological considerations of scientific knowledge into account, we often do a poor job of communicating the relationships of science to other forms of knowledge, preferring instead to portray science as a distinct knowledge form within disciplinary boundaries that perpetuate a synthetic distinction between science and other disciplines, and school knowledge and knowledge 'of the real world'. What school is for, is another question that will be attended to in this book. As alluded to earlier, we cannot be satisfied with school as part of a technocratic machinery that churns out 'literate' individuals much as a factory will churn out widgets to standardised specifications ready to use in machines. Yet at the same time, limiting what is to be learnt to the idiosyncratic nature of problems that will afflict societies from time to time robs the students of the potential for learning the types of knowledge that have been termed 'powerful' by sociologists of curriculum. It is these knowledges that are to be counted as cultural achievements of humanity, that provide ways of thinking that may transcend time and space, to find use in contexts distant from its sites of generation. Scientific knowledge tends to fall into this category; and so, a careful balance needs to be struck, between two extremes of a reverent and uncritical scientism and an overly parochial and unnecessarily utilitarian approach.

Advances in understanding *how* we learn has also provided us with an additional support for learning through the context of the makerspace. Even if we are to only concern ourselves with the cognitive goals for science education, and not bother with the social liberatory, or even cultural developmental goals, there are good reasons to consider what has traditionally been called 'learning by doing'. The philosophical movement of anti-intellectualism, not to be confused with the desire of some to elect political leaders that they can quaff alcoholic beverages with, suggests to us that there is something about knowing beyond what can be represented; we can know more than we can tell. It has only been the recent several decades where research in the cognitive sciences turned their attention to what has now been termed as the project of embodied cognition, considering how the body is an essential element of the cognitive apparatus. These perspectives are slowly finding their way into studies of education, and this new understanding of learning will no doubt be useful in thinking about why makerspaces are useful sites for science and innovativeness. The chapters in this book are therefore organised as mentioned below.

In the next chapter, I will begin by discussing perspectives from the sociology of curriculum. A long-standing argument concerning the nature of knowledge concerns the status of knowledge claims—quite simply, while it is clear that knowledge can confer power to its possessors, there is controversy over whether this power arises out of its intrinsic value or due to the fact that the knowledge is produced and has limited circulation among people who are already in possession of power. A useful distinction, used by scholars such as Michael Young, is between knowledges of the powerful and powerful knowledges. Earlier in his career, Young advocated, as with many others, for a position of social constructivism: that truth claims were essentially social constructions, and as a result, there was a certain arbitrariness (and hence injustice) in the selection of particular forms of knowledge deemed important to communicate through schooling. A clear example of this may be the choice of readings in English literature: why should students in, say, rural Indonesia read only Shakespeare instead of local English authors? This position was vastly aided by developments in STS, especially of the second wave, who took it upon themselves to demystify the nature of knowledge generation in science and technology, to 'peer behind the curtain of the sorcerer's stage', so to speak, and to uncover a weak, hypocritical man (it was always men) caught up in his own webs of deceit. If knowledge claims were essentially up to the discretion of people in power, there was a social justice purpose in advocating for the opening up of the curriculum to more diverse ways of knowing, and bringing in knowledge claims of those whose voices have previously been deleted. In more recent years, there has been a gradual realisation that the nature of knowledge claims may be more nuanced than that: while some claims are truly up to the tastes of powerful groups, some others are made in reference to an empirical reality which, at the very least, can serve as a close approximation to a neutral arbiter for a large class of truth claims. Some truth claims may even transcend the time and space of its generation, as when centuries-old mathematical problems are solved with contemporary insights, and produce applications to cryptographic communications unimaginable then. The question for education becomes further complicated when we understand that the production of these truth claims often consists of procedures that are not amenable to representation. While enough knowledge of physics (say) can be acquired to understand and perhaps critique public knowledge claims, the ability to produce new knowledge requires tacit knowledge which cannot necessarily be communicated, only acquired through performance. Similarly, while traditional disciplinary boundaries provide culturally accepted optimal performative means of acquiring these tacit knowledges, interdisciplinary approaches especially of comparably nascent fields such as design and particular forms of engineering can pose powerful knowledges which may not have had time for adequate representational formulation. The problem here is one of untangling the mess of truth claims: as curriculum scholars, we necessarily want to communicate to students' claims that are true. However, what constitutes truth, and what may be some heuristic methods for the generation of truth claims that we ought to communicate to students? If particular approaches to knowledge generation appear to be tacit performances that are not particularly amenable to representation, where does this leave us on our dichotomy of powerful knowledges and knowledges of the powerful? Could it be that there exist powerful knowledges which *are* knowledges of the powerful? What implications do these deliberations have for curriculum?

Chapter 3 continues with the question of tacit knowledge by more deeply investigating the nature of this form of knowledge. The main premise here is that we learn using our entire bodies and the environments we live in and that 'head knowledge' is essentially incomplete. For instance, the classic example attributed to Michael Polanyi of riding a bicycle presents a case of knowing more than we can tell. Just because we cannot explain how it is, we respond using the rotational inertia of the front wheel to balance out the tendency to fall in a particular direction, does not mean that we are not in possession of a kind of knowledge about it. If we consider the established disciplines such as the natural sciences, particular representative devices have been invented over the years to deal with the problem of communication among peers. Much of the difficulty with acquiring knowledge in the disciplines is in the specialised language used by practitioners: either the terms acquire meanings distinct from pedestrian usage or specialised terms are invented altogether. Considering abstract mathematical operations, for instance, whose meanings have been popularly thought of as free-floating concepts not tethered to any form of reality, we come to understand from the work of people such as Rafael Núñez that it is possible to trace the meanings we place on such operations to bodily experiences that the human species possess in common. For instance, because matter is conserved, we can abstract basic arithmetic from our day-to-day experiences of interactions with matter. And, as our bodies are inevitably embedded in particular contexts, particular cultures, and their specific ways of thinking and communicating about the experiences we encounter, these cultural framings are important sources of our knowledge about the world. It is important that as educators, we become more clear about the connection between the experience and the representation, and the risks that arise out of severing such connections. It is entirely possible, to use a metaphor attributed to John Dewey, that we can confuse the map with the terrain; we can attempt to teach the representations without reference to the bodily experiences that ground these representations. Given the wild claims of the impending or existing 'information age', it can be rather easy to suppose that learning can occur simply by interactions with a screen. The purpose of this chapter is to minimally cast doubt on this project and to consider what we are missing when we leave out the body in learning.

Chapter 4 introduces the science educator to design as a way to think about the kinds of goals that makerspaces can aspire towards. Design, as a late entrant to the academy, has not had as much scholarly attention as the other disciplines. Also, unlike other disciplines, it is a 'poorly behaved' discipline in that the boundaries of its expertise are not sharply defined against other disciplines. Consider how various groups working in fields as diverse as clothing fashion, nuclear power plant construction, and website user experience all can claim to be designers. Or that, all these diverse ways of enacting design actually do share a common set of working principles that can be introduced to students as a productive way of thinking. In schools, for the sake of introducing material that is simple and clear, we often strip out unnecessary details and, at some level, even to the point of insulting their intelligence when school problem sets involve little more than extracting the relevant information to 'plug into' equations that they are supposed to remember. Problems that occur in real life almost never appear to us neatly announcing the kind of ways we are supposed to solve it. Significantly, there is a special class of problems that deserve our particular attention: termed as wicked problems, these problems are non-trivial, and practically all problems that involve humans are wicked to some degree. Wickedness in this context does not mean evil, but rather refer to a special form of intractability to do with not really being able to pin down, once and for all, what an ideal state ought to be. I will explore the design method as a particularly suitable method for responding to some of the concerns brought up in the earlier parts of this book. In brief, if we desire students to make use of the knowledge that they acquire in schools for a public good, an understanding of science is not enough. The problem lies with the ends to which we put scientific knowledge towards: what counts as a public good is not necessarily well defined. To make matters worse, the philosophical device of Hume's Guillotine reminds us that regardless of the accuracy in our description of the current state of affairs, these descriptions have no ethical bearing on any prescription that we care to provide. The natural sciences need to be recognised for its role as the creation of very accurate descriptions of nature, and how we can predict the future behaviour of the things we study. However, scientific knowledge cannot offer us any particular prescriptions forward; especially in matters of social beneficence. The best knowledge we have on matters of prescription belongs to the humanities, and the particular discipline that deals regularly with a purposeful transdisciplinary intersection is design. The purpose of this chapter is to review some of the knowledge bases about what design knowledge constitutes, what benefits design confers to its users, and how teaching design in school contexts can reveal some shortcomings in the way that school as conventionally structured can contribute to the project of 'future proofing' fully participating citizens of tomorrow. This chapter will also discuss how making, innovation, and the natural sciences are related to design and how an alternate vision for schooling may look like. We will discuss the nature of design problems and point out the wicked problem as a major characteristic of design problems. Considering a selected subset, wicked problems: (i) do not have once-and-for-all solutions, (ii) are ill-structured, and (iii) have solutions which create further problems. Immediately, it is not hard to recognise that the problem of schooling is fundamentally wicked in nature. How should, for instance, one approach the instruction of design in schools? Its transdisciplinary nature often means that design requires students to draw from diverse sources of knowing, yet design knowledge is not a linear superposition of its parts. While schools tend to value known standards, often with highly quantitative measures, good design requires a diversity of qualitatively distinct goals often impossible to prespecify. Design requires an authentic assessment of user needs, whereas schools tend to restrict students to synthetic contexts. If we are to be serious about schooling as a form of preparation for participative citizenship, it would be beneficial to apprehend design accurately, due in no small part to the wicked nature of sociopolitical problems. We contend that schools can and should serve as 'rehearsal space', with students using design to inform their increasing participation in society, and with authentic problem contexts providing meaningful challenges for the acquisition of both disciplinary knowledge and the creative dispositions required to provide valuable solutions to the world.

Chapter 5 expands on the theme that was begun in Chapter 4. In recognising that science is a useful mode of analysis, but must remain mute on issues of what ought to be made real, it has been nonetheless an amazing sleight of hand that society has collectively given these experts carte blanche to decide on the kinds of futures we ought to have. Certainly, contention about regulation is evidence that societal mechanisms do exist to try to rein in the excesses of the possible harms that accompany the use of particular scientific methods and technologies. However, a significant problem is that regulation often lags innovation, by which time large amounts of money have been invested in these 'disruptive' ideas, and organisations built around them would have become 'too big to fail'. Consider, for instance, the experiments in social media that have been thrust upon society. Built around a model of widespread access paid for by advertisers, these organisations have become multi-billion dollar behemoths for which regulation can be stymied by greed amplified with financial resources. What does not help matters is the seemingly mass illusion of technological determinism—that scientific knowledge and technologies are derived from considerations outside of the society and, when sprung onto society, are capable of reshaping society as we know it. Such a belief underlies our very common practice of referring to historical periods in terms of the dominant technologies of its time—from the stone age, through the various metals, into industrial revolutions, the fourth of which we are apparently still undergoing. Much of the rhetoric about the existing threats to society seem to rely on an as-yet undeveloped technology to save the day; we seem overly confident that technology will have all the answers. Yet, as we have seen from Chapter 4, the risk that technological solutions will create side effects of its own is actually fairly high. A science and technology education that does not address technological determinism only serves to further misrepresent it, further deepening the many crises that seem to be amplified by our use and misuse of science and technology. Besides technological determinism as a social orientation to technology that can mislead, a question that needs to be asked is whether science and technology, in and of themselves, may possess particular intellectual orientations that may be inappropriate outside of its practice. Here, the problem can be summed up by the popular aphorism—that if the only tool one possesses is a hammer, every solution starts to look like a nail. If we only have the intellectual orientations of science and technology, what sorts of problems might result? Consider the curious correlation of engineering talent and militant Islamic Jihadis: among militant Islamists, there is an unusually high concentration of STEM talent that cannot be explained by (social) network effects or the demands of complexity of the improvised explosive devices that they favour.

It seems that the kinds of training in STEM, with the tendency to see the world in absolutist, black-and-white terms, and the tools that amplify intentions to 'set the world right', can be exactly what is needed to recruit extremists. Certainly, this must be an extreme example (pun not intended), most of us who know STEM are not especially prone to violent overthrow of power. But it still remains that the enlightenment ideal is the ability of science to speak truth to power, and if we were to educate students with these intellectual orientations in mind, what sorts of problems might we expect societies to suffer as a consequence? Might it be that our reliance on binary true/false statements makes us susceptible to misinformation especially when truth lies in the fuzzy boundaries of 'it depends'? Especially on normative matters, it should be clear that no one position has a monopoly on truth. With these ideas in mind, the question this chapter engages with is that of how science should be taught, and especially, what may be the affordances of makerspaces in attending to these issues?

In Chapter 6, we want to think about the nature of our material interactions with reality. While scientific progress is admired for its deep insights, we can often lose sight of the material interactions that help deliver these insights to researchers. Working in separate fields, Tim Ingold and Andrew Pickering arrived at similar conclusions that help inform how we ought to organise learning. Quite loosely, while we think of materials as passive and merely receptive to our idealised designs, closer analysis reveals that materials respond to our machinations, behave in often non-deterministic manner, and offer resistances to human agency almost as if it has an agency of its own. While Ingold arrives at this conclusion from study of Art, Architecture, Anthropology, and Archaeology, Pickering's approach came from studies of the SSK production. Ingold takes an anti-hylomorphic stance, arguing that even for a discipline such as architecture where we conventionally believe plans to be laid down with finality before the first bricks are laid, modifications to the intent of the architect occur all the time. Hylomorphism, or the belief that artefacts are the result of abstract forms superimposed onto matter, can be thought of as a relatively intuitive approach to making. Seen as a cousin to beliefs about the dualist nature of consciousness, it should be no surprise that hylomorphism has been around for a long time. However, as with intuitive physics or other 'cognitive modules' that we are thought to be imbued with to survive over evolutionary timescales, much of these intuitions are just plain wrong. While in science instruction, we occupy our time with such abstract concepts as frictionless surfaces, d-orbitals, or Krebs cycles, getting an experiment to work to reveal these idealisations often involves a great deal of manipulation. This is not to deny or denigrate the ontological status of theoretical objects but rather to point out that the translation between theory and reality can be tenuous indeed. More than that, getting material reality to respond in ways congruent to theoretical predictions can be difficult indeed. This difficulty does not arise depending on investigators' competence or lack thereof but rather from the nature of materials in and of themselves. For example, Pickering points to the best scientific effort possible to contain the Mississippi river as it threatens to inundate New Orleans—despite the most accurate hydrological data and physical modelling available to the Army Corps of Engineers, the river regularly defies the expectations of the various devices put in place to direct the river. While a more 'scientific' description of the phenomena would ascribe this behaviour to the complex nature of river flows, Pickering (and peripherally, Ingold) would argue that it may serve as a kind of intellectual shorthand to consider these non-human factors to have a form of agency. Crucially, this buys us a different nature of scientific knowledge: we need to consider, as Pickering suggests, that our knowledge is at best, *correct*, and not necessarily *true*, and that any usage of scientific knowledge requires users to be competent in a performative aspect that is a superset of its representational aspect. This position, not incidentally, appears to overlap with the anti-intellectual approach of Chapter 2: scientists *performing* science can know more than they can *represent*. This position requires us to consider carefully how schooling ought to communicate scientific knowledge.

Finally, in Chapter 7, I want to bring all the themes together to consider how (science) education can be an agent for social change. If the foregoing chapters portray science education within the context of makerspaces as a possible means of creating a different future for schools, there is the obvious question of why this book is needed. Why would this vision of education not already be the standard for schools? For me, a large part of the problem lies with the conservative nature of the institution of schooling, with the dominant approach being that of the recitation model. When new models for school appear, the tendency is to try to incorporate the new through the interpretive lenses of the old. This can result in what some have termed as a cargo cult complex, where there can be an outward appearance of change, but no significant difference in terms of how people think about the novelty. For the past several decades, researchers have documented how many grand sounding technological innovations promised to 'revolutionise' schools, only to notice later that very little has changed. How might we avoid the pitfalls of these kinds of promised revolutions? It seems like the main thing that needs to happen is the slow, unglamourous change of cultural taken-for-granted notions of what school constitutes. To do so, we need to think carefully about what schools are supposed to be good for. For the most part, schools tend to be associated with economic preparation of individuals (and, therefore, the community). In my opinion, this needs to be rethought. Yes, economic preparation is important, but it cannot be all there is to school. Schools can do a lot more, and science education in particular need not only concern itself with 'conceptual acquisition', and only at a cognitive psychological perspective. Science education can, and should, consider itself as a source of moral uplift, and I will discuss three principles of makerspace instruction that can help bring this about.

★ ★ ★

To be sure, this book is a fairly ambitious project that draws from a wide enough range of sources that it is likely to leave some experts frustrated at the lack of depth at the individual topics covered. In my opinion, this may be an unavoidable consequence of the nature of schooling: by its very complexity, the stereotypical version of a scientific approach of isolating variables and manipulating individual variables is unlikely to 'work' in the same way that one may conceivably expect medical sciences to work. For one, as Labaree [cite] writes, unlike in medicine where patients' cooperation is not always necessary for treatment, education requires willing participants (the alternative is *indoctrination*). And, in any reasonably authentic context for educative practice, all possible 'layers' of analyses will intermingle to produce confounding effects, from genetics (even though educators will avoid this layer), individual cognition, all the way through to sociological and political ways of understanding the world. To make a complex situation worse, the demands that educational practice *be* educative opens up the discussion to ethical considerations for which the best intentions may not necessarily result in the best outcomes. It is this ethical dimension that drives the 'what for' in the title of this book, and which consideration of what we ought to do with making as a learning activity needs to be clear about. It should be clear from this chapter that my position is for a science education that is socially beneficial, even though appreciating what makes something beneficial necessarily brings us away from the natural sciences, and can be unsettling to educators used to clearly demarcated boundaries between the traditional school subjects.

Note

1 After Arthur C. Clarke's aphorism that 'any sufficiently developed technology is indistinguishable from magic'.

2

WHAT IS KNOWLEDGE ANYWAY? OBTAINING CLARITY ON KNOWLEDGE AND ITS ROLE IN SOCIETY

This is a book about makerspaces, broadly conceived as a site for science, technology, engineering, and mathematics (STEM) education. While it can be easy to think of STEM knowledge as well-established, easily catalogued collections of factual knowledge that all students need to have some grounding in, and that makerspaces merely the contemporary equivalent of science teaching laboratories for the 21st century, it would be a mistake to do so. Just as science educators in the last century have contended with the role of the science laboratory as a venue for science learning (DeBoer, 1991), a parallel problem for this century may be the role of makerspaces as a site for STEM education. Similarly, just as much of the arguments about science education in the past century hinged upon advances in our understanding of the nature of science, it is important for us to become clearer about the nature of STEM knowledges and its implications for education. In this respect, this chapter recounts the major shifts in ontological and epistemological thinking about the nature of science in the past century, as a means to outline the major positions of thought about the nature of knowledge. As we shall see, these positions are somewhat limited, and they result in different positions as responses to the question of what and how we should educate. This chapter will go into some detail of the theoretical movements in the nature of science, as it is rather important to get a sense of the relationship between knowledge, societies, and the role of schools in its dissemination.

It is perhaps convenient to illustrate the problem with a contemporary example. At the time of this writing, we have been continually reminded for the past several years now that we live in 'post-truth' times, that particular leaders wish to be able to assert truth by fiat; in science-related news, anti-vaccine 'activists' claim a link between vaccines and autism, women have been targeted for sales of dubious 'health' products hawked by a famous actress, and in at least some school systems in the United States, evolution by natural selection still remains

DOI: 10.4324/9781351116220-2

a contentious issue. Social media and the ability of the internet to allow all and sundry to broadcast their ideas and opinions to the rest of the world have been accompanied by all manner of 'fake news', falsehoods circulated several times around the world even before 'the truth has time to put on its pants'.[1] At the heart of this problem is the nature of truth—what is it? What influences the truth value of a claim? We are often used to quoting authoritative sources: influential newspapers, textbooks, trusted leaders; yet all the above cases point to a central problem for us: can we separate truth claims from the identities of the source. More precisely, the question is one of the separability of knowledge and power (Biesta, 2013; Foucault, 1977): can truth claims ever be independent of the power invested in its source? To trust that something happened in a particular manner because it was reported in one particular manner is to trust that the source is truthful; more often than not, the warrant for trusting a source derives from its past record of truthfulness. Yet, we have been confronted in the last half century or so with an intellectual approach that has sought to dismiss truth claims as either a privilege of the powerful, or irrelevant in comparison to the effects such claims may have on the world (Lynch, 2004).

These issues are relevant to STEM and makerspaces because, in large part, we desire education to be communicating truth to our students. While many will reach to concepts in the established sciences such as those governing motion, chemical reactions, and biological systems as 'established facts' and the least likely to be up for contention, this impulse reveals a rather uninformed understanding of the nature of science. While these knowledges are durable, in that they are not susceptible to change in the coming decades, they are nonetheless tentative, as when relativistic mechanics subsumed classical mechanics, acid–base theories got updated, and DNA became known to be the store of genetic information. How these knowledges can get updated and how students can come to be part of the process of discovery and knowledge production should be important aspects of the STEM curriculum. In considering the interdisciplinary challenge posed by STEM, it is also important to distinguish between the kinds of knowledge claims. Loosely as an introduction, it should be clear that the knowledge of the sciences and mathematics are slightly more durable than that of engineering and (especially) technology: for instance, if a major corporation declares that a particular computing language or engineering method will be important in the coming decade, it should not be hard to spot the potential conflict of interest. It is easy to see that almost all the biggest issues that confront our societies and ways of life have a scientific or technological component—it is after all science and technology that have enlarged the scope of our abilities. For better or worse, the *djinn* has been released, and there is no way to put it back in again; what remains is for us to learn to make use of our knowledge in the best way possible.

Yet, that very term can (and should) cause us at least some hesitation: what, exactly, constitutes the best way; and how would we judge anyway? With powerful interests vested in particular outcomes, it can become unclear for whom these decisions are being made. In schools, we take it that at least one of the purposes of schooling should be communication of a vision for the better life. It is not unexpected therefore that we often look upon the current problems and at least implicitly expect schools to be part of the solution. Proposals have been made, for instance, amidst grand claims of the death of the university as we know it, to centre curriculum around particular grand problems that confront society at this point in time (Taylor, 2009). Such proposals are often accompanied by pronouncements of the arbitrary and counterproductive nature of disciplinary boundaries, often in less flattering terms. Predictably, a somewhat conservative position argues for the intrinsic nature of these knowledge boundaries [cite]; that the distinction between different forms of knowledge is not synthetic and actually refers to different ways in which phenomena present themselves to our understanding. This is no easy problem, of course—what characterises true truth claims that we ought to communicate to students? How much of truth claims are obligatory, and are the way they are because of the nature of reality, and how much of it is the result of social negotiation, and therefore a consequence of a (possibly unfair) distribution of sociopolitical power? In other words, borrowing the terminology of sociologists of curriculum Michael Young (Young, 2008a) and others, is it that the powerful become so because they have acquired particular knowledges, or is it that the knowledges used by particular groups are only powerful because they are associated with the powerful?

In considering makerspaces as sites of inter- or even trans-disciplinary forms of learning, these problems should be matters of serious consideration for scholars. In fact, this is easily the core problematic for curriculum, expressed in the simple dictum offered to all beginning students of curriculum: what knowledge is of most worth; or modified with the sociological perspective to become: *whose* knowledge is of most worth? In makerspace specific research, a main focus appears to have been in demonstrating the educative benefit of making activity, often via numerical measures of desired outcomes. However, less well established is the curriculum perspective, for example, somewhat wistfully expressed in a recent journal article: "It looks like they are having fun, but *what* are they learning?" (Bevan, Gutwill, Petrich, & Wilkinson, 2015, emphasis added). Makerspaces are conventionally associated with such technologies as digital fabrication and rapid prototyping, and particular approaches to design; in all cases we can always turn on the critical perspective and ask: why learn about, say, the Arduino microcontroller platform? Why code in python, and not C++? Just because certain design methods have been effective to generating solutions for one group, need not mean that these methods ought to be universal. If we consider the methods used by, say, leading corporations in the design and manufacture of handheld computers with integrated telecommunications technologies (yes, the smartphone), do we say that we ought to promulgate their methods because they are in possession of some novel and uniquely effective methods at 'innovation', or should we be more sceptical and say that their methods often only work because of the vast sums of money that support their processes? With technology companies loudly advocating 'coding' as a school subject, do educators have any well-considered response to these fashions, or is it that the educator role is to simply be the expert in communicating whatever it is that capital interests want? What, then, is the expertise of curriculum scholars and school board specialists?

Clearly, we need to gain some deeper insights into the nature of knowledge, to have a clearer idea of the characteristics of educationally beneficial forms of knowledge, and for whose interests acquiring these knowledges serve. One particularly difficult case that has captured the attention of numerous scholars over the years happens to be that of science. While its history extends literally to the first human to ever wonder about the natural universe and consider if there was regularity to the apparent chaos, it is probably more useful here to discuss a more interrupted history starting our retelling from the last 130 years or so. Around this time, what needed accounting for was the spectacular successes of science and technology in understanding and changing the world. Physicists were making so much progress in understanding the nature of the physical universe that there was actually some concern that all of physics was soon going to be 'solved'. We begin our retelling at around that time.

What makes science special?

There is a character in a British sketch comedy routine called *Little Britain*, where a character Carol Beer, played by comedian David Walliams in drag, is the customer service representative at a travel agency. In these skit segments, customers present requests, to which Carol, in a highly reluctant and contemptuous manner, will respond by haphazardly banging on her keyboard, and in a deadpan note, almost always goes 'Computer says no'. While the point of the skit is to poke fun at the typically unhelpful attitudes of people in customer service, what is of more interest to us here is how much of the situation we recognise and therefore find hilarious. Why is it that we often tolerate being told by a computer, or a related form of technology for that matter, what we can, or cannot do? Why do we defer to the people who have specialised access to these technologies and 'scientific methods', as in Carol in this example? Carol is designed to make us laugh at ourselves when we recognise that there can be bad actors hiding behind machines who summarily refuse our requests when there is not even the intention of trying. If we scale this problem up, this very same kind of scepticism with people hiding behind machines, technologies, and obscure methods is ultimately the same sentiment that leads people to climate change denialism. While it is easy to claim that 'more science' will lead more people to understand just what is going on, and hence a greater acceptance of the consensus position, the underlying problem is nonetheless an important one. We live in societies based on specialisation of roles, and we cannot expect everyone to understand how, for instance, to repair a drowned smartphone. We have to be able to trust that when Carol says the computer says no, there is somehow a good reason for the request to be truly impossible. Yet, we often know, as with Carol, that it is more often than not a human or an organisational decision that causes this refusal.

The tension, then, is between the politics of the organisational intent and the possibility that the knowledge claims being put forth are in fact accurate.

The central question that has occupied much of scholars in science and technology studies (STS) has been the epistemological position of science and technology—what is it that makes the knowledge of science and technology special, and its considerations override so much of our realms of existence? As a simple shorthand for understanding the field's intellectual history, it appears that we have gone through approximately three waves of thought, with the first two being the two extremes possible on this issue, and the third one we are currently undergoing interested in a hybrid position. Respectively, the first two movements or waves have been called positivism and social constructivism. There is as yet little consensus on what this third wave ought to be dubbed, but some theorists have referred to this movement as social realism [cite], or simply as the third wave of STS [cite Collins and Evans]. Some scientists have held rather deprecating attitudes towards the study of science, with Richard Feynman famously having been quoted as saying 'Philosophy of science is about as useful to scientists as ornithology is to birds'. Such a position is inaccurate and misrepresents what scientists need to know. To a less explicit extent, scientists need to know, among other things, what constitutes acceptable forms of evidence, what are the social mechanisms for judging the veracity of truth claims, and the purposes of science, all of which are concerns of the philosophy of science. In any case, in order to communicate scientific knowledge accurately, it is likely that educators require the perspective of the 'ornithologist' rather than the bird—by analogy, if we are trying to teach fledglings how to fly, it is more than likely that adult birds have forgotten the characteristics of flight and how upward flight differs from, say, gliding. Similarly, the educational concern has been with the nature of science, or NOS, but largely in the way that knowledge of NOS conveys a more accurate depiction of science. Less well studied are the educational implications of the anthropological concern of the role of human behaviour in the scientific enterprise. Specifically, these range from the psychological to sociological concerns (deep ornithology, if you like), of (for instance) how truth claims are negotiated through empirical investigations with material reality, to how sociocultural factors determine what science gets done, and what ought to be the interests of science. This is, without a doubt, a deliberate transdisciplinary practice—if knowledge of scientific concepts alone is insufficient, neither is just psychology or sociology. A major consequence of this is that the anthropological and philosophical approach to understanding the production of scientific truth claims is not going to be easy. Let's begin with an abbreviated history of ideas in this field.

If we care about truth claims, the first thing we may need to develop is a sense of scepticism about the entire venture—if we begin with the question of how is it we can have *any* form of truth claim at all, we notice, as with epistemologists, that there are three basic modes of justification. We could have an infinite regress of justifications, which should be considered an unacceptable form of justification. That leaves us with two satisfactory options: justifications based on (apparently) self-evident truths or justifications based on a chain of mutually reinforcing claims. Respectively, these strategies have been termed foundationalism and coherentism (Williams, 2001) and recognised since the time of the ancient Greek philosopher Agrippa (hence this is called Agrippa's Trilemma) as the only means for justification. As I will try to show below, these strategies map onto those that have been used to describe scientific justification; understanding the strengths and limitations of these approaches will explain curriculum and instructional strategies that have resulted. The general strategy here is to explain the historical precedence of educational approaches and then propose that our current approach needs to be updated in recognition of our contemporary understanding of the nature and philosophy of science.

Positivism

In early attempts to understand the special place of scientific knowledge, an obvious starting point was in the fact that science concerned itself with empirically grounded truth claims. Due to the apparent successes of science in modernising all dimensions of life, scientific and technological forms of organisation were perceived as the ideal. To a great extent, we are still held in the sway of such a perspective when we obediently defer to the Carol Beers in our lives. The positivist perspective is essentially the epistemological strategy of foundationalism—that there exist empirical referents to ground truth claims, that there can be reliable and stable relationships between an independent reality and the truth claims that we may make of it. On its face, such a strategy does not seem problematic, it would appear to most people that when we talk about such things as mountains, trees, animals, and their properties, these objects exist and do not require the presence of humans and our cultural forms of representations to give it a sense of actuality. Such an approach also appears unproblematic when we consider the astounding successes of science and the results of the application of such forms of knowledge. Positivism was the first response to the problem of explaining the unreasonable success of the natural sciences. In its historical context, such a question was completely meaningful: around the turn of the 20th century, great improvements were being made to the standard of living. There were great strides in understanding physics, to the point that many believed that physics was a solved problem, and that all that remained was the tying up of loose ends [cite]. The industrial revolution had run for over a hundred years, creating great wealth in western countries; improvements in public health allowed more people to enjoy their new found wealth for longer periods of time, and the stability of the western ideological viewpoint made it tempting to theorise that all forms of life and social organisation needed to be arranged in the same rationalised manner of the natural sciences. This was also the time of 'scientific management' [cite] and the Taylor rationale for curriculum [cite]; the idea being that one could replicate the success of science by using its methods.

Of course, these 'methods' were a poor facsimile of what scientists were actually performing; *the* scientific method was eventually shown to be anything else but methodical. Even today, the roots of the desire to ground truth claims in empirical evidence run deep, in education research textbooks it is fairly routine to open with a chapter about the methodological or science wars that have dominated discussions about the validity and reliability of one method or another. Almost predictably, and to no surprise to anybody even vaguely familiar with the caricatured 'scientific method', a cause of controversy has been the centrality of quantitative methods as the prime method of the natural sciences, and the appropriateness of attempts to extend this form of reasoning to other domains of experience and understanding. A key distinction that may help bring light to this conflict is the distinction between topological and typological variation (Lemke, 1999). From this point of view, in order to account for variation in the world, one can consider that there are two main forms that we are capable of describing it. In topological variation, we are interested in the *degree* to which two quantities can be said to differ. For instance, if we are interested in how much hotter a cup of coffee is than another cup of tea, we are interested in the topological variation of the property of heat. On other hand, the typological variation is interested in the difference between the various properties that make the cups distinct, such that one can tell coffee from tea. As should be obvious, quantitative measures are ideally suited for describing topological variation, *once standards can be agreed upon*. On the other hand, numerical methods do less well for describing typological variation, where qualitative descriptors become necessary. Could we describe, say, the distinction between the phenomena of Brexit and the election of Trump by numerical means? What numbers would we choose to aid our attempts at description, and what could that even mean? The quantisation process—the conversion of selected qualities into numbers that can then be compared or have patterns noted—can understandably be a controversial process.

The choice of standard measures to use to describe variation can appear to be a matter of social convention and as such a completely arbitrary process. It was only recently that standards of measurement of physical quantities became rationalised and made dependent on natural quantities; nonetheless, even in the definition of the standard metre, for instance, we detect an arbitrariness: the metre is defined as exactly the distance travelled by the speed of light in $1/299,792,458$ of a second. This is in an effort to preserve the currently agreed upon but complete arbitrary definition of a metre, traditionally referring to a length of metal kept in a vault in France. There is no empirically obligatory reason why a metre should be the ideal unit, and not, say, the imperial unit of a foot. The outline of this form of argument, taken further, will eventually bring us to the next section where we discuss social constructivism, whose essential premise can be surmised (badly) as the casting of doubt on the scientific project by pointing out the socially agreed upon nature of truth claims.

Typological variation remains yet another hard problem. Consider the scientific attempts to categorise natural variation: it is well understood that attempts to define species by reference to typological characteristics are at best a tenuous process. The lack of clearly demarcated boundaries between species can mean that these attempts at classification and producing scientific claims (Havstad, 2011) can further contribute to the belief that the scientific venture is based on shaky epistemological foundations. So far, then, the prospects for attempts to demarcate science from non-science don't seem particularly good, in that while there could be a plausible case to be made for the strong empirical basis for truth claims made in the natural sciences, exactly what is the relationship between such claims and the natural phenomena is the subject of contention—do these truth claims have an obligatory nature, and refer to something that actually exist in the real world, or are they matters of social convention? And that is even before we begin the debate on the nature of the 'real world'—is there even such a thing? While most of us would recoil at the thought of the non-existence of the 'stuff of life', the situation is rather more difficult for theoretically 'proven' objects such as electrons, double-bonded carbon atoms, and vaccines; and at the other end of the scale, plate tectonics, evolutionary timescales, and globular superclusters.

The implication for educators ought to be clear: as educators, we ought to be communicating justifiably true statements; knowledge, not merely information, or worse, mistruths. While it may have been initially intended as a means to communicate the high status knowledge of the natural sciences and technology as a means to open up access to learners particular ways of seeing the world, this problem of the arbitrary nature of some truth claims began to haunt researchers of education in the earlier half of the 20th century. The matter, as may be elegantly surmised by sociologists such as Michael Young (2008a), is a conflict between the knowledges of the powerful on the one hand and powerful knowledges on the other. To be sure, there is a certain degree to which this conflict is not a choice between binaries; nonetheless, it remains an important consideration for educators, specifically scholars of curriculum. With special relevance to the examples that confront us in our time, the methods of computer science, data analytics, and psychologically profiled targeted advertising (among others) raise issues about the status of these forms of knowledge. Should one communicate these knowledges to students in the hope that they may be able to obtain access to the appropriate corridors of power and obtain the 'better life' (of some sort)? If, for some reason, the large personal data harvesters' business models become legislated out of existence, could our students trained in the erstwhile valuable methods be able to use their now obsolete expertise in other pursuits? Before we explore these implications further, let us return somewhat to the problem of the status of scientific knowledge as a signal case about these issues.

Social constructivism

At some retelling, the rise of social constructivism, social constructionism, or allied methods whose aim was to cast doubt on the obligatory nature of scientific project coincided with the post-war state of mind. If the natural sciences could be used to, in a sense, support the wars of chemistry and physics that were the World Wars I and II, respectively; if the natural sciences and its associated technologies (including the early information technologies developed by the IBM (Black, 2001)) could be instrumental in the most efficient ethnic cleansing operation, something was obviously quite wrong. The bright shiny dreams promised by the ever progressing development and spread of rational thought represented by the natural sciences gave way to fears of mutually assured destruction and nightmare scenarios such as the cobalt bomb (Aftergood & Kelly, 2002). It became worryingly apparent that the natural sciences could be used for evil as well as the good that seemed to have been promised by the earlier, more innocent times. This is not to say that pre-war science was completely directed to social betterment. For instance, Stephen Jay Gould's (1981) history of craniometry, physiognomy, and early intelligence studies is a signal lesson on how the best science of its day could be used to respond to the utterly grotesque social impulse of 'proving' the superiority of the European 'race' over others, and men over women.

The 'first shot' of the contemporary science wars may have been fired by Karl Popper; when he rather earnestly started questioning the status of scientific knowledge, pointing out the problem with inductivism. Popper wanted to interrogate the growth of scientific knowledge, and how it was that we could have truth claims in the sciences. Essentially, the accepted method of justification of knowledge in the positivistic interpretation of the sciences took an inductive approach: As the 'pile' of evidence mounted in the direction of one particular position or another, one could be increasingly sure of that favoured position. Inductivism had its doubters due to the essentially open-ended nature of proof: one cannot prove for all time and space that, say, Newton's theory of gravitation was correct throughout the universe. It was Popper's insight that falsification held the key: that there was a crucial asymmetry between verification and falsification. While hypotheses cannot be proven true with inductive accretion of knowledge because knowledge boundaries are essentially unknown, we can always accept a hypothesis that has an in-principle possibility of being proven false, but which has yet to be. In a sense, Popper's work heralded an age of curiosity about the nature of truth claims in the sciences; if the question was how scientific knowledge developed and grew, the next response of Thomas Kuhn broke open the walls to the castle of science. Kuhn (1962/1996) argued, among other things, that there are somewhat arbitrary reasons why science progresses from one 'paradigm' to another, that such movements produce 'incommensurable' truth claims across the chasm of the paradigms, and that much of the rationale for the adoption of one paradigm over another could be attributed to something akin to an intellectual fashion.

Coming as it did in the time when the humanities was becoming concerned with the so-called linguistic turn towards the analysis of language use and the slipperiness of meaning, Kuhn's ideas were widely received by many groups. Specifically, for philosophers, social scientists, and researchers concerned with the methods of exclusion of selected groups from the 'grand narratives' of Progress and the larger social compact, Kuhn's work began signalling the centrality of the sociological lens to research efforts in the epistemology of scientific knowledge. While Kuhn did not himself use sociological methods, the kinds of queries that were suggested by a reading of his work certainly existed. For instance, it is possible to interpret that Kuhn suggested that scientific progress depended on factors beyond the intellectual and that moments of knowledge progression had more to do with social agreement than the fact that current paradigms were incomplete:

> Both normal science and revolutions are, however, com–munity-based activities. To discover and analyze them, one must first unravel the changing community structure of the sciences over time. A paradigm governs, in the first instance, not a subject matter but rather a group of practitioners. Any study of paradigm-directed or of paradigm-shattering research must begin by locating the responsible group or groups.
>
> *(Kuhn, 1962/1996, p. 179)*

Kuhn also argued for what he termed the *incommensurability* of paradigms that there was no essential progress towards truth via cumulative accretion of knowledge and that truth claims made under one framework was at least in principle not intelligible under another. For instance, under Kuhn's interpretation, Einstein's theories about motion cannot be said to be an improvement over Newton's; they are merely different. More serious adherents may even claim that the Newtonian conception of mass refers to a different theoretical object than Einstein's, and these two concepts serve different purposes within each of their frameworks. Headline grabbing was Kuhn's insistence that these purposes were incommensurable: someone working within Newton's theory would not be to explain their conclusions using Einstein's and vice versa.

Not long after, actual studies of the practices of scientists as social collectives emerged, detailing the sociological conditions under which scientists practiced in order to make the truth claims that they were making. Especially prominent were studies by Bruno Latour (1986, 1993). His work centred on portraying the work of science and scientists as 'just' another form of inquiry, equivalent to other disciplines and occupying no special access to truth. The motivation for such studies was an inversion of sorts—that while we may have (by that time) extensive studies of faraway 'primitive' tribes, less is known about a particular tribe embedded into our culture that nonetheless provides much cause for celebration and concern in our daily lives. The cultures of the practitioners of science, that produce wholly remarkable claims about the world that we live in, still remain mysterious to this day: the fact that many people were convinced of conspiratorial wrongdoing during the so-called *Climategate* episode should be evidence that there is insufficient public understanding of the way scientific evidence is gathered, and how scientific claims are made. Latour studied the members of the Salk Institute with the same ethnographic interest as one would an alien culture, not necessarily taking scientists' (often after-the-fact) recounting of the processes of discovery, but instead studying the process as it happened.

Just as sceptics suggested during the *Climategate* scandal, Latour discovered that the pursuit of scientific knowledge is just as mundane as any other human venture. This is not to say that scientific knowledge does not deserve any special epistemic status, but that the process of its production was not more special than others. Scientists were prone to wishful thinking, much of the work that goes on in science was pretty mundane, and *'The'* scientific method was but a myth. In fact, Paul Feyerabend (1975/1993) made the claim that he is now most famously misunderstood for, that 'anything goes' in the pursuit of scientific knowledge. It was around this time that the academy fashionably started to claim that, among other things, power and knowledge were essentially inseparable (Foucault, 1977), that all knowledge was socially constructed [cite], and that because we cannot become clear about the influence of power on knowledge, all forms of knowledge are essentially suspect. If this sounds very much like the opening sequence for the post-truth world, we currently inhabit, it is in no small part because there are elements of truth to it. While the precise relationship between post-modernist relativistic excesses and the current post-truth politics may be a matter for significant academic debate (Collins, Evans, & Weinel, 2017; Lynch, 2017; Sismondo, 2017), the outline of the charge is rather clear: this process of dismissing truth claims by referring to the latent political interests supposedly behind each and every truth claim has spread from the academy outward into the larger society (Kakutani, 2018). If sociologists could 'unmask' vested interests (Hacking, 1999), especially in the supposedly neutral ground of the natural sciences, there was no essential barrier to using these methods for truth claims in other fields with a greater degree of obvious bias. Political discourse quickly became a fraught affair, and matters of science that needed public consultation started to break down, as it became far too easy to point to vested interests like big pharmaceutical companies that stood to gain from compulsory vaccination.

Social realism

If positivism and social constructivism could be considered the first two waves that characterised how researchers and the public came to understand the epistemology of science in the 20th century, it is perhaps inevitable that the third wave that followed charted an intermediate course between what could be considered its two predecessor extreme positions. Laudan (1996) argued that an overly strict interpretation of previous positions' intellectual demands caused an unnecessary error of reasoning. As explained by Laudan, the central problem with the philosophy of science was the relationship between rival theoretical interpretations of the phenomena and the phenomena in itself. As language is necessarily limited and limiting, the question arises of how we should adjudicate between these rivals—which one is true, or perhaps, closer to the truth of the phenomena; which one gives a better description? Of course, such a framing is possibly problematic: what *better* means depends a lot on who is asking the question. We often consider that better theories are superior because they provide more technically accurate descriptions of phenomena or that a more contemporary theory needs to be able to encapsulate an earlier theory in terms of explanatory extent. Yet, this is not necessarily the case, as scientists have been known to favour theories which are analytically 'more elegant' [cite]. Geocentrism, which has been superseded in the public imagination by heliocentrism, is no less accurate in its predictions; heliocentrism is merely deemed as more elegant.

In Laudan's interpretation, the linguistic turn beginning around the 1930s began to cement the view that there was no such thing as language that was independent of theory and that by the time of Kuhn's writing, the concept of incommensurability of theoretical viewpoints became an easily accepted notion:

> In more than a figurative sense, postpositivists turned scientific theory comparison into cultural anthropology, and objective theory comparison became a casualty of cultural relativism. Even if the anthropologist herself can eventually come to assimilate the alien culture and language, there is no way she can communicate them to her fellows. In the same vein, Kuhn insisted that scientists who have 'gone native' with respect to a new paradigm cannot explain it to their less venturesome colleagues. Rival theories became, not merely different worldviews, but different worlds, different realities.
>
> *(Laudan, 1996, p. 9)*

This almost ideological commitment to the incommensurability of rival theories did the postpositivists no favours: the dominant conception was that because theories could not be compared due to the incommensurability thesis, *all* theories were therefore doomed to a cognitive relativism, that factors other than the intrinsic quality of theories (*viz.*, social relations) were the determining factor. The interpretation of theories shifted from its supposed superiority as determined by scientists, to now a more open and public examination of the aesthetics and axiological standards held by these scientists. In a sense, the public trust of scientists fell dramatically from its postwar high of heroes whose inventions (such as radar or the atomic bomb) helped 'us' win the war, towards the beginnings of distrust during the postwar population boom period where people started to question the horrors that may arise from a misapplication of science, all the way through to the post-truth times we find ourselves today.

Social realism arose as a response to the excesses of the anti-science critique. The basic position of social realism is a rejection of the hard binary positions advocated by both the positivists and the post-positivists. If the positivists believed that truth existed 'out there' and that we may obtain objectively true knowledge, the post-positivists insisted that such a truth was not available because of the tainted nature of adjudicating between rival claims. Because human notions of values and beauty, and not merely objective criteria such as the correspondence of truth claims to reality was used to judge theories, this meant that *all* of science was suspect, so claimed the postpositivists. Noted commentators such as Feyerabend (1975/1993) went so far as to dismiss the possibility that there was any real method or systematicity to science in itself, drawing from historical cases of outstanding personalities that 'anything goes'. These were not the outlying cranks and frauds, but apparently, respected scientists such as Galileo or Copernicus, people whom science textbooks typically held up as paragons of the scientific method.

Laudan's response (1996) to Feyerabend's general critique was to firstly note that his historical retelling may not have been entirely accurate. Even if the recounting was accurate, the fact of individual scientists' rule breaking does not warrant his extreme response; Laudan writes:

> But Feyerabend's inferential leaps from the historical record are a good deal grander than I have thus far suggested. For his ultimate strategy is to argue that-because he has discredited certain modern-day methodologies of science (especially those associated with Carnap and Popper)—we are thus forced to conclude that the whole enterprise of delineating the rules of scientific methodology is ill conceived. This is a monumental non sequitur. It is rather as if one argued that, since most scientific theories eventually fail, it follows that we should stop doing theoretical science altogether (p.104).

In a similar vein, Collins and Evans (2017) surmise that there have been three 'waves' of Science Technology Society (STS) studies, corresponding to positivism, postpositivism, and social realism as I have recounted above. The basic position of social realists is the assertion that reality really exists: 'if you can spray them they are real' (Hacking, 1983). While social processes are responsible for making and adjudicating truth claims about the nature of reality as we may perceive it with our senses or with instruments, and while such social processes may be fallible because individual humans are fallible, the net effect of these processes is to produce a self-correcting mechanism that, over the long term, corrects itself and represents humanity's best knowledge.

So what?

If the foregoing has been a somewhat unexciting excursion into the philosophy of science, it may be worthwhile asking ourselves at this point what these ideas buy us. Here, I wish to bring forth the educational implications of these ideas, especially since many of these writers propose grand changes to social institutions demanded by their insights. Feyerabend proposed that societies needed to be defended *against* science:

> For example, consider the role science now plays in education. Scientific 'facts' are taught at a very early age and in the very same manner in which

religious 'facts' were taught only a century ago. There is no attempt to waken the critical abilities of the pupil so that he may be able to see things in perspective.

(Feyerabend, 1975, p. 4)

Similarly, compare such calls with public misunderstandings of the workings of science that led to a widespread belief of a Soviet era disinformation campaign alleging that the Human Immunodeficiency Virus (HIV) was an artificial virus engineered by the US Central Intelligence Agency which somehow spread uncontrollably. More recently and in the similar vein concerning human health, spurious connections between vaccines and autism have been drawn; climate change denialism still threatens populations; people have been terrified online regarding the rampant deaths linked to dihydrogen monoxide; and the list goes on. It is probably too easy to suppose that the solution to these problems is 'more science'. Almost predictably, curriculum committees think about how scientists and science literate people become the way they are because of interest. Therefore, means must be put in place to enhance the interest and engagement of populations, to 'get them when they are young'. Hence, public school teachers are saddled with the responsibility of getting young people excited about science.

Clearly, we have tried this before, and it does not work as well as we imagine it to. It is probably not more knowledge of scientific facts that can guard against public misperceptions of scientific principles. Just as there will be people who will never appreciate theatre productions, there will be those for whom scientific principles will not appeal. Given the social nature of many contemporary scientific issues, and the diversity of opinions we should want to cater for, how should public decisions be made regarding the deployment of scientific and technological inventions that could simultaneously threaten our collective well-being at the same time as it enhances our lives? As science educators, we should guard against both extremes of an uncritical scientism and a radical distrust of science. These issues, minimally, present curriculum problems, requiring us to consider what it is that we want students to learn. As signal cases that represent the major epistemological positions, students should understand not just the end results of scientific inquiry, but also the processes through which we may arrive at these positions.

To be sure, this argument is not new; Richard Duschl (2008) what he termed as a 'three part harmony': balancing demands of the conceptual, epistemic, and social learning goals as an approach to move past science education as a means of selecting future scientists. This chapter expands upon the epistemic goals of science education, providing an overview of the major epistemic positions that have formed the backdrop of the intellectual positions over the years. The main purpose for rehearsing these arguments is as to create a defensible rationale for the curriculum of a science, STEM, or makerspace programme. As I asserted in the opening of this chapter, we want such a curriculum to be true, to offer superior knowledge that will be of utility to students in uncertain futures. Yet, as we can see, how we conceive of epistemic justification principles can make us more or less susceptible to particular forms of knowledge.

Just as a shortcut to considering curriculum knowledge, sociologist of curriculum Michael Young (2008a, 2008b) started his early career critiquing the degree to which power was vested in the school curriculum. Knowledge, he had claimed, was a product of arbitrary socially negotiated process, and as such, school knowledge posed problems for social justice because it privileged knowledge of particular classes of people—notably rich, white, men. It should not be a surprise to learn this position gained much traction within the academic community in the height of the post-positivist movement in the 1970s. Forward 40 years, and the same, albeit more senior, academic issues a mea culpa, acknowledging that there are limits to which knowledge can be arbitrarily made up; and that there exists what he now termed 'powerful knowledges' [cite]. Such knowledge has obligatory qualities in that the relationship between the truth claim and a reality which exists is limited by *the way it is*. Crucially, possession of such powerful knowledges can provide individuals with access to sites where decisions are made about important aspects of life, and communicating these powerful knowledges can serve an important social justice purpose. We may want school knowledge to be enjoyable and be relatable to the people who learn it, but to insist that this be done at the expense of learning the powerful knowledges may not be a good idea. As with Young, I propose that it is important to distinguish the knowledge we want to communicate, from the means that we may wish to do so. We need to distinguish between curriculum (*what* ought we know) and pedagogy (*how* we should teach it).

Of course, nothing in life can ever be so uncomplicated: there can be situations where pedagogy is irretrievably confounded with curriculum; particular orientations to inquiry, for instance, are almost completely communicated through the manner of the pedagogy. What is significant for us to consider here is the almost unquestioning manner in which makerspaces are associated with particular forms of curriculum and pedagogy. The emblematic devices of the 3D printer, microcontroller, and discrete electronics make the assumption that makerspaces are about digital fabrication, the design method, and a style of pedagogy that is associated with a Californian/American mindset (not to necessarily imply anything negative here). Minimally, this book sets out to trouble these automatic associations. What is the curriculum for makerspaces, for STEM, for science education that seeks to bring a more constructive element? What ties makerspaces, STEM and making activities together? What are worthwhile curriculum goals for making? Pedagogical approaches? While we appear no closer at this point in responding conclusively to these questions, we can move forward with useful ideas that can buy as much theoretical sophistication. If we accept the social realist epistemological position, we already begin to see the importance of making for a complete STEM/science education. If our knowledge is in the form of socially generated truth claims grounded in stable reality, it matters that we increase access to the experiences of this stable reality and that learners have opportunities to negotiate truth claims made up from their subjective experiencing of this reality. Key to this assertion is the notion that not every experience can be adequately represented, made into 'knowledge'. To expand on this insight, I have to move on to the next chapter.

Note

1 A quote, interestingly, often *mis*attributed to Winston Churchill. Its origins are unclear, but has been traced even to as long ago as the early 1700s.

References

Aftergood, S., & Kelly, H. (2002). Dirty bombs: Response to a threat. *Journal of the Federation of American Scientists, 55*(2), 1–11.

Bevan, B., Gutwill, J. P., Petrich, M., & Wilkinson, K. (2015). Learning through STEM-rich tinkering: Findings from a jointly negotiated research project taken up in practice. *Science Education, 99*(1), 98–120.

Biesta, G. (2013). Interrupting the politics of learning. *Power and Education, 5*(1), 4–15.

Black, E. (2001). *IBM and the holocaust: The strategic alliance between Nazi Germany and America's most powerful corporation* (1st ed.). New York, NY: Random House Inc.

Collins, H., & Evans, R. (2017). *Why democracies need science*. Cambridge, UK: John Wiley & Sons.

Collins, H., Evans, R., & Weinel, M. (2017). STS as science or politics? *Social Studies of Science, 47*(4), 580–586.

DeBoer, G. E. (1991). *A history of ideas in science education: Implications for practice*. New York, NY: Teachers' College Press.

Duschl, R. (2008). Science education in three-part harmony: Balancing conceptual, epistemic, and social learning goals. *Review of Research in Education, 32*, 268–291.

Feyerabend, P. (1975). How to defend society against science. *Radical Philosophy*, 11 (Summer issue), 3–8.

Feyarabend, P. (1993). *Against method* (3rd ed.). London: Verso. (Original work published 1975.)

Foucault, M. (1977). In A. Sheridian (Ed.), *Discipline and punish: The birth of the prison*. New York, NY: Random House.

Gould, S. J. (1981). *The mismeasure of man* (1st ed., 352 p.). New York, NY: Norton.

Hacking, I. (1983). *Representing and intervening: Introductory topics in the philosophy of natural science*. New York, NY: Cambridge University Press.

Hacking, I. (1999). *The social construction of what?* Cambridge, MA: Harvard University Press.

Havstad, J. C. (2011). Problems for natural selection as a mechanism. *Philosophy of Science, 78*(3), 512–523.

Kakutani, M. (2018, July 14). The death of truth: how we gave up on facts and ended up with Trump. *The Guardian*. Retrieved from http://www.theguardian.com/books/2018/jul/14/the-death-of-truth-how-we-gave-up-on-facts-and-ended-up-with-trump.

Kuhn, T. S. (1996). *The structure of scientific revolutions* (3rd ed.). Chicago, IL: University of Chicago Press. (Original work published 1962.)

Latour, B. (1986). *Laboratory life: The construction of scientific facts*. Princeton, NJ: Princeton University Press.

Latour, B. (1993). *We have never been modern (p. ix, 157 p)*. Cambridge, MA: Harvard University Press.

Laudan, L. (1996). *Beyond positivism and relativism: Theory, method, and evidence*. Boulder, CO: Westview Press.

Lemke, J. L. (1999). Typological and topological meaning in diagnostic discourse. *Discourse Processes, 27*(2), 173–185.

Lynch, M. P. (2004). *True to life: Why truth matters*. Cambridge, MA: MIT Press.

Lynch, M. (2017). STS, symmetry and post-truth. *Social Studies of Science, 47*(4), 593–599.

Sismondo, S. (2017). Post-truth? *Social Studies of Science, 47*(1), 3–6.

Taylor, M. C. (2009, April). End the university as we know it. Retrieved July 2018, from http://www.nytimes.com/2009/04/27/opinion/27taylor.html.

Williams, M. (2001). *Problems of knowledge*. Oxford: Oxford University Press.

Young, M. (2008a). *Bringing knowledge back in: From social constructivism to social realism in the sociology of education*. New York, NY: Routledge.

Young, M. (2008b). From constructivism to realism in the sociology of the curriculum. *Review of Research in Education, 32*(1), 1–28.

3

AN ANTI-INTELLECTUAL APPROACH TO KNOWLEDGE AND LEARNING

If the previous chapter gives us a brief recounting of the intellectual history of the epistemology of science for the past century or so, there is an important respect in which the retelling is incomplete. As the title of this chapter indicates, there is an anti-intellectual perspective which is missing. In opposition to the previous chapter where we are more concerned with a social, public communication of representations of phenomena, in this chapter I am more concerned with its subjective, private, and individualistic *experiencing*. If the motivation for the previous chapter was to derive some curriculum principles for the determination of what it is we should know, this chapter is more concerned with the ways in which we can come to know, and how these subjective experiences are related to the more public forms of knowledge. In this chapter, I will set out to consider the following questions: What does it mean to know? How are representations related to phenomena? What instructional implications exist for learning in makerspaces and especially for the nurturance of innovation? To outline the argument, I find it useful to defer to an age old paradox that warrants an occasional retelling, if only to remind us that despite the advances in our understanding of how the mind works, we may have come not much further in our understanding of how we may learn new things. Ultimately, the satisfactory resolution of this paradox involves the invocation of partial knowledge, a concept that needs further elucidation as to its nature and its instructional implications. One candidate concept to perform the explanatory duty of partial knowledge is that of embodied cognition, which I will review and derive some educational implications of.

What is a makerspace?

In consideration of the previous chapter considering the curriculum goals of science education, and this chapter considering the latest insights into how we think and learn, it is now appropriate to consider what constitutes a makerspace

DOI: 10.4324/9781351116220-3

and what form learning activities ought to take in such a space. The previous chapter provides us with some curriculum principles—makerspaces, if considered sites for science education, ought to facilitate the communication of the uniqueness of scientific knowledge, for instance. To be sure, makerspaces have more recently been associated with the growing science, technology, engineering, and mathematics (STEM) movement, to which many others have added to the acronym the arts (STEAM) or even the humanities in general (SHTEM). The goal of this book, in part, is to work towards a consideration of these diverse versions of curriculum goals for makerspaces. However, at the outset, it may be simpler to consider the first letter of the acronym, which may have outsize influence over the rest of the letters.

STEM and makerspaces have shared parallel success over the last few years, judging by the numerous trade and research publications. Google scholar returns over 10,000 results for 'makerspaces' and over 3 million for 'STEM education'. Makerspaces, judging by the more popular publications, has been perceived as sites for the grounding, or the application of abstract scientific principles, and for increasing student engagement with STEM concepts. While these are good goals for makerspaces, there is a certain degree to which these goals serve as low hanging fruit. In many ways, this is not unexpected. The confluence of several factors, most notably in the United States, has made some of these goals more likely. It has been a recurrent trend in the United States (and many other capitalist societies around the world) discourse in education to suggest that education needs to contribute to the development of economistic and even national security goals.[1] For instance, historians of science education easily point to the Space Race of the 1960s caused by the panic of the orbiting Soviet Sputnik satellite, the 1980s 'unilateral economic disarmament' against the economic threat posed by the ascendant East Asia 'Tiger economies' (DeBoer, 1991). More recently, concerns that manufacturing and 'STEM' professions, and the associated economic edge conferred to societies which possessed such talent, were being lost in the global offshoring of manufacturing (Martín-Páez, Aguilera, Perales-Palacios, & Vílchez-González, 2019).

This, and the proliferation of contemporary technologies such as digital design, digital fabrication, and tiny, embeddable computing platforms, have coincided with the associated rise of 'educational technology' to see the makerspace increase in importance in recent years. On top of that, the accessibility of these technology platforms has been accelerated by their relative low cost. Due in part to the expiry of patents associated with 3D printing, for instance, there are now consumer digital printers which promise to make 'push button manufacturing' a reality. Small microcontrollers like the Arduino and the BBC micro:bit can be purchased for less than $30; these latter devices and their associated free software platforms mean that basic robotics and general programmable behaviour can be designed into objects fairly inexpensively. Perhaps most importantly, the rise of 'user generated content' on public fora, and social media monetisation, has led to the growth of instructional and community sites online which promise 'free' lessons and access to a community of 'free' experts to serve as mentors for all manner of projects.

These approaches to learning, and making, make an implicit assumption that there are, or least can be, two distinct phases of the process. Much like how an increasing number of products are being labelled as 'Designed in X, Manufactured in Y', there is a widespread perception that it is possible, or in fact desirable, to distinguish the 'thinking' from the 'doing'. As any person who attempts cooking recipes from online tutorial videos would have the experience of, following recipes or instructions is not at all a simple affair. The heart of the matter is the problem of information reduction and abstraction—how much information is necessary to express the subjective experiencing of the phenomenon? If I wish to communicate the experience of kneading bread dough to one who has not tried it before, it will be fairly obvious that the written word will be found deficient compared to a video, and the video will present less information than actually being beside a 'live' demonstration. Extrapolating, this would necessarily mean that the best thing to do when learning how to make something is to experience it for oneself. Yet, on a daily basis in schools all over the world, we communicate the reverse; we privilege the abstract representation, the word, the design, over the making.

To belabour the point, there are even book length treatments into the apparent asymmetry. We believe in the primacy of the 'head' over the 'hands' that the greasy mechanic is not as 'smart' as the suited academic. From one who has actually done both, Matthew Crawford (2009) points out the deeply cognitive nature of tinkering and manual troubleshooting when it comes to, for instance, trying to figure out why an aged motorcycle does not want to start; and conversely, the stupefying experience of being a hired pen to write justifications for policies already decided. For us in education, the problem we face is to consider the nature of learning—is it a more a matter of communicating abstract representations, for which the recitation model of schooling is probably sufficient? Or is it more a matter of communicating the subjective experiencing of phenomena, for which an apprenticeship model would be more appropriate? In many jurisdictions, the conventional logic is that of the mind-hands divide, that scholars do not really need to use their hands, and mechanics do not really need to use their minds (or we design systems such that there *becomes* a separation of thinking from doing). Such a position may be termed as intellectualism, for its privileging of intellectual ability and the abstract representation over the messy, gritty realities.

Given the scientific propensity for reductionism and the quest for elegant and all-encompassing 'theories of everything', it can be easy for educators to get swept up in the logic of intellectual development at the cost of a certain sense of disconnect: we may be developing excellent symbolic manipulators who have no idea what the symbols actually mean. Drawing from the philosophical thought experiment (how ironic!) of Searle's Chinese Room (Searle, 1980), we need to be deeply dissatisfied with the notion that students who have competence with the symbols actually understand the meaning of the symbols that they are working with. Conventionally, we seek students who are competent in the 'application' of these knowledge, as if it were a simple matter of being able to deduce pertinent aspects of these theories to the particular contexts of their deployment. In the last 30 or so years, an increasingly well-supported position argues for the reverse, that abstractions are not enough, especially for learners acquiring competence; this position may be described as an anti-intellectual approach for its valuing of experiences over representations. As can be deduced, the goal of this chapter is to prepare a case for makerspaces as sites for this kind of anti-intellectual form of learning. Makerspaces could, and should, become sites for a new kind of learning experience, if only because we now know more about the way we learn.

Meno's paradox

So, the story goes that Socrates was asked by the younger Meno if virtues can be taught. Meno recounts his teachers' principles that virtues are specific to individuals, depending on their position in society, whereas Socrates seeks a more universal set of virtues that are applicable to all. Eventually, the argument progresses to the question of how one may recognise a virtue; this was an especially acute problem with Socrates' method of inquiry that was predicated on his ignorance. He sought knowledge by first acknowledging his ignorance and then asking questions of everybody in order to fill the void. Meno argued that this method contained an inherent paradox: how could Socrates know when he came across a response that was helpful in resolving his ignorance? If he knew what he was looking for, then there was no need to have gone searching. On the other hand, if he did not know what he was looking for, the answer could be staring at him in his face, and he would not know it. A contemporary Socrates would be having the similar problem, except with search engines and an internet full of sites deliberately out to misinform. Considering the numbers of individuals who have 'dropped out' of mainstream understanding about the efficacy and safety of vaccines, or the sphericity of our planet, it is quite obvious that Meno's paradox still is a relevant problem today. And this is not to suggest that it is only a problem for isolated cranks in far flung corners of the internet; as many of us with well-intentioned elderly relatives warning us about the hazards of cancer causing microwaved foods would know; this problem is one of limited 'scientific literacy'. I use the term in quotes to denote that the key is not quite so much the possession of a huge database of facts, but in the possession of a particular orientation to discerning whether or not claims to truth have any value. This problem is not about the possession of facts, again, because the problem is with attempting to adjudicate the truth value of a *new* claim.

Meno's paradox is essentially an epistemic paradox; one of several, these paradoxes provide us with a means to question our taken-for-granted notions that we can have straightforward access to knowledge and its means of production. Here, the paradox presents us with the challenge of learning and incorporating new knowledge into our existing conceptual frameworks, especially if we are to make use of methods of inquiry and are not simply told from authoritative sources. Intuitively, we see that this paradox is resolved fairly easily because our knowledge does not have a binary quality between complete ignorance and full comprehension. We can 'sort of' know something, we can have partial knowledge, and we can judge veracity by its correspondence to other truth claims that we are already in possession of. However, what is interesting for this chapter, and the book as a whole, and what I intend to expand on in this chapter, is the nature of this form of partial knowledge. What does it mean to 'sort of' know something? How might partial knowledge influence strategies for teaching and learning, especially in makerspaces, for STEM learning objectives, and for innovation?

Making tacit knowledge more explicit

One entry point to the notion of partial knowledge is that of tacit knowledge. Perhaps most famously attributed to Michael Polanyi, whose pithy phrase 'we can know more than we can tell' (1966/2009, p. 4) opening many a journal article or book chapter. Also famous in Polanyi's characterisation of the problem is that of riding a bicycle: without being able to say how we do it, we can know *how* to ride one. The essential problem here is one of attempting to understand the connection between the wholly private and subjective realm of the experiencing of phenomena, and the socially agreed upon 'objective' means by which we may represent and communicate our subjective experiencing to others (Collins, 2010). Tacit knowledge appears to be an entry point for our understanding of the anti-intellectual approach to learning. Here, an explanation is probably warranted: when we speak of learning, intellectual development is often considered an ideal goal. However, the goal of such an anti-intellectual approach is not an avowal of intellectual development, but rather a privileging of the experience first, before its abstract representation.

Such is the stance of Abrahamson and Kapur (2018), Fyfe, McNeil, Son, and Goldstone (2014), Nathan (2012), and Nathan et al. (2014). For example, Nathan's position is that while formalisms (the term he uses to refer to abstract representations) have been powerful for their ability to 'reify tacit knowledge, highlight deep structure, and retain and transmit cultural knowledge' (p. 126), we have to distinguish between what is good for the discipline with what is good for the learner. Far too often, we often deploy a formalism-first perspective for learning, largely because experts who have familiarity with the power of these conceptual tools believe that students need to have competence with them. But, as Nathan surmises, this does not necessarily mean that the best way for learners to arrive at such competence is to *begin* with the formalism. Nathan points out that the roots of this line of thinking lie deep within the Western philosophical tradition, with Plato as one of the original sponsors of the attempt to derive purely abstract principles that would be unchanging and eternal, compared to the material realm of 'blooming, buzzing confusion'. Nathan also notes that 'the elite of the Hellenic and Greco-Roman eras held the use-oriented engineer and technologist in contempt' and that 'manual labor was relegated to those of lower social status, often slaves'. While little has changed in our thinking about the relative status of the 'pure' and 'applied' subjects in the intervening 2000 or so years, a growing branch of education and psychological research has found it meaningful to upset this 'formalism first' perspective and to give a more meaningful, more empirical grounding for the concept of tacit knowledge. Current research is focussed on ascertaining the degree to which contextual factors may provide distractions away from the acquisition of abstract conceptions, and it appears that a growing amount of findings points towards the importance of concrete experiences first (see, e.g., the special issue of *Instructional Science* edited by Abrahamson and Kapur, 2018).

To be sure, the psychological concept of 'discovery learning' somewhat associated with this approach is not new, and meaningful objections have been made based on the grounds that 'real' reality can pose distracting detail which may delay the development of conceptual schemas (see, e.g., Kirschner, Sweller, & Clark, 2006; Mayer, 2004). After all, as the aphorism concerning scientific theory development goes: observations without theory are blind, theory without observations is misguided. Based on the recurrent argument for the growth of scientific knowledge, contending between an inductive-accretion versus a hypothetico-deductive model (e.g., Clement, 1989; Lawson, 2005, 2010), the problem is one of the diverse possibilities for the particular form of information we may want to consider as signal, as opposed to noise. One position would argue: without some form of framing, how would we even know what to look out for? This, again, echoes Meno's paradox in the search for truth. However, the point here is not to deliberate on the process of creation of new scientific knowledge (this will be done in a later chapter), but to point out certain long-running arguments that will have some influence on how we may think about the problem of makerspaces, innovation, and science education. If we can acquire implications to curriculum from sociocultural considerations of the nature of knowledge, the ambition here is to derive pedagogical implications from the psychological and cognitive. A contemporary solution to the problem of learning may derive from a consideration of embodied cognition.

Embodied cognition

In thinking about thinking, a predominant line of thinking has been that there exists something immaterial, which is responsible for consciousness. Entire religions have been built upon the notion of an immaterial soul that persists even after the body has expired. Following the scientific norm of only desiring natural explanations to phenomena, we see that this position, commonly referred to as Cartesian dualism, is untenable: not only are we left with supernatural factors whose existence we have to take on faith, Descartes postulated the existence of what was termed the 'Cartesian witness', the sole audience member of the 'Cartesian theatre' (Dennett, 1991), the place where cognition all comes together. This creates a problem of regress—if we can explain consciousness by means of a witness, how could we then explain what goes on within the witness. By 'exorcising' the soul from scientific explanations of consciousness and cognitive function, we are forced to develop new metaphors for understanding the working of the mind. Given the rise of information processing theory and the rise of computation in the last century or so, it is perhaps inevitable that recent models that have success involve the mind as a computer.

Perhaps our best explanation to date derives from the computational theory of mind (CTM) (Pinker, 1997, 2002). Derived fundamentally from a 'strange inversion' (Dennett, 2009) in reasoning credited to Alan Turing, the CTM (Fodor, 1975) asserts that in order to make a machine that could compute, it was not necessary that this machine understands anything approaching basic arithmetic. We take this idea for granted these days, trusting our computers—dumb little hunks of semiconductors, metal, and plastic—to bring entire planeloads of people across large distances and perform other contemporary miracles, but it is also this same principle that has guided research in cognitive sciences in the last half century (Horst, n.d.). The CTM allows us a means to open up the black box of the mind and to exorcise what Gilbert Ryle called the 'ghost in the machine'. Recalling world war era 'computers' which were banks of humans working on subroutines of complicated mathematical problems, the CTM explains the workings of the mind by successively decomposing computational tasks into smaller and smaller pieces which structures within the brain could act upon in some non-magical way. Indeed, much of artificial intelligence (AI) research could be seen as a practical verification of these developments in our understanding of mental processes (Dennett, 1998).

In thinking about the computational problem, AI researchers came across what was called the frame problem (Dennett, 1998). In designing robot interactions, a persistent problem was the resolution of factors deemed pertinent to the task at hand—exactly what should the intelligence take for granted as factors pertinent to the processing of the task at hand? While this problem was eventually solved for AI implementations, Dennett points out that this remains a new epistemological problem, similar, but not quite identical, to the problem of induction: How do we know what aspects of a problem are pertinent to its solution? In considering the problem of mind, we come to an analogous problem: how can we be certain that computation resides solely in the brain? It becomes obvious to ask this question when we think of using such devices as calendars, abacuses, or even simple pen and paper as we work out long division. Perhaps, as Whitehead surmises, thinking is hard:

> It is a profoundly erroneous truism [...] that we should cultivate the habit of thinking of what we are doing. The precise opposite is the case. Civilization advances by extending the number of operations we can perform without thinking about them. Operations of thought are like cavalry

charges in a battle—they are strictly limited in number, they require fresh horses, and they must only be made at decisive moments.

(Whitehead, 1911, in Rowlands, 2010)

Researchers surmise that just as we offload and extend our sensory perceptions with tools like callipers, microscopes, gas chromatographs, we certainly can offload computation to 'devices' (generally speaking) outside of the physical brain. In many ways, we are, as Clark (2003) surmises, natural born cyborgs, 'forever driven to create, co-opt, annex, and exploit nonbiological props and scaffoldings'. In effect, the embodied mind thesis takes the rejection of the Cartesian dualism to its logical end—not just rejecting the ghostly immateriality as explanation for mind, but also similarly rejecting the notion that the brain as the only organ of cognition. An analogy to the circulatory system is probably apt here: while the heart is a central organ of the circulatory system, the functions of the system cannot be met by the heart alone—the blood vessels, while literally peripheral, are essential parts of the system, as anyone who has suffered a stroke would know too well.

If we must consider that cognition takes place not only in the brain, what may be some principled means of theorizing and interpreting what is and is not cognition. One famous approach, offered by Clark and Chalmers (1998), essentially argues that:

> If, as we confront some task, a part of the world functions as a process which, were it done in the head, we would have no hesitation in recognizing as part of the cognitive process, then that part of the world is (so we claim) part of the cognitive process. Cognitive processes ain't (all) in the head!

For example, as Clark and Chalmers surmise, we would have little difficulty in accepting that the process of using a piece of paper to record a note to oneself for later recall is functionally equivalent to storing the same note in some form of mental memory. With minor modification, however, Rowlands (2010) points out that the emphasis of the Embodied Mind Thesis (EMT) is not in the equivalence of cognitive states with environmental structures:

> The thesis of the extended mind should not be understood as claiming that cognitive states can be identical with environmental structures. Properly understood, the thesis makes no claim about cognitive states at all. It is a thesis that concerns cognitive processes and it claims some of these processes are, in part, composed of processes of manipulating, exploiting, or transforming environmental structures. It is the things we do with external structures—our manipulation, exploitation, and transformation of them—that constitute properly cognitive parts of overall processes of cognition (p. 67).

This point needs to be belaboured here: it is not as though that, when we do long division on paper (say), the inscriptions on paper are identical to the mental states that we possess as we perform the cognition. Rather, it is the process of manipulating abstract symbols referring to particular quantities that are identical to cognitive processes, if they were to be completely done 'in the mind'.

If this position is correct, then the implications for education are interesting: if cognitive processes are embodied, why should our instruction be exclusively in the modalities of sight and sound? If kinaesthetic experiences are essential to our understanding of abstract scientific principles, surely the traditional class-room emphases on sitting passively are mistaken. In mathematics education, a recent special issue in the *Journal of the Learning Sciences* featured precisely these perspectives; in learning about complex numbers, Nemirovsky and associates (2012) found that gestures and perceptuo-motor activity were essential for the communication of mathematical insights; Alibali and Nathan (2012) make a case for embodied learning of mathematical concepts through studying the gestures of teachers and students. Bergen and Feldman (2008) also point to many studies which link perceptuo-motor functioning to cognition, most notably in under-standing figurative and abstract meaning, where, for example, concepts are more quickly understood if sentences simulate certain more plausible scenarios. For example, 'The road runs through the desert' is understood more quickly than a sentence like 'The fence climbs up to the house' because one mentally simulates the speed with which one can move along the described paths:

> All these lines of research point to a common conclusion. Conceptual proc-esses make use of the internal execution of imagery, qualitatively similar to the past experiences it is created or recreated from. As such, using concepts is quali-tatively similar in some ways to experiencing the real-world scenarios they are built from (p. 318).

A more complete review may be found in the work of Lawrence Barsalou (2008, 2010, 2016), outstanding findings include: cognitive simulations of past embodied experiences increase perceptual fluency and the likelihood that per-ceptions are categorized correctly; cognitive processing is heavily dependent on simulating the motor actions underlying it; individual differences in spatial abil-ity correlate with the ability to draw inferences; and, cognitive simulations play a central role in abstract reasoning.

The embodied perspective is perhaps best demonstrated by the work of Lakoff and Johnson (1999) and Lakoff and Núñez (2000). In the former, Lakoff and Johnson make the case that we come to understand abstract thought mostly through metaphorical extensions of embodied experiences. For instance, the 'CATEGORIES ARE CONTAINERS' mental schema utilises the sensorimo-tor domain of spatial awareness based on the primary experience that things that go together tend to be in the same bounded region. In the latter, Lakoff and Núñez explain mathematical reasoning by reference to embodied schemas, building metaphor upon metaphors to generate meaning for such concepts as irrational numbers, transcendental numbers, and, eventually, Euler's famous identity: $e^{i\pi} + 1 = 0$. While the authors leave the educational implications open for interpretation, we can at least infer that concrete manipulatives and other embodied experiences are important, not just for children in the early stages of development, but also perhaps for learners of all ages. More than that, these findings point to an obvious chasm in our understanding of how these embodied experiences affect learning: what constitutes an educative embodied experience? How can teachers best utilise embodied experiences in learners' acquisition of abstract conceptual ideas? Are there any regularities in learners' responses to embodied experiences that teachers can exploit to instructional advantage?

Makerspaces and the future of labour

Makerspaces can go beyond their current typical use case as sites for introducing neophytes to new technologies (as in science museums) or as sites for the technology equivalent of 'science lab' sessions where students 'apply' material learnt during lectures. The embodied cognition perspective suggests to us that grounded experiences and gestures are important aspects and mechanisms of learning. If linguistic metaphors and mathematical abstractions ultimately derive their meaning from embodied experiences, it would stand to reason that the larger the library of embodied experiences we can offer to students, the greater the potential for students to develop abstract metaphorical extensions based on these experiences. Makerspaces need to be perceived as sites where these grounded experiences are acquired. Such experiences can be developed, likely in a non-deterministic manner, through metaphorical extensions into abstract theoretical representations.

While these rationales for makerspaces and practical activity of a particular sort—construction and engineering particular configurations to solve problems—may provide a technicist justification for its use in learning, there is a sense that such a rationale may ultimately be found lacking. The position here is one of a profound dissatisfaction with forms of 'educational technology' whose aim is to ensure a mechanistic relationship between the 'inputs' and the 'outputs' of the educational interaction. Here, technology is interpreted widely, and not limited to digital computing and networked devices, to also include cultural technology such as the mass lecture which needs to be considered as a somewhat unnatural method of communication. Besides the potential for some of the claims of efficacy or utility of educational technology to be overstated to the point of being 'bullshit' (Selwyn, 2016), there can be insufficient thought to the intrinsic educative value of particular interactions in and of itself. It is a subtle argument—that we should use means to educate that are educative. To use a somewhat more lurid metaphor, taken from a poster held up during an anti-war march, I had the opportunity to witness some years ago: bombing for peace is like raping for love. Just because we understand how the mind works and how learning takes place does not mean that we may use these knowledge in ways to coerce and control learning without the agentic participation of our students. Just because we know, for instance, how to use gamification technologies to cause addiction does not mean we should design addictive games that ostensibly teach students about high-value schooling content. I will have more to add to this idea as the book develops, but the general idea I borrow from Gert Biesta (2016), that the core of the educative interaction is one of risk: not from the possibility that teachers and students do not try hard enough or are not sufficiently ready for the interaction, but from the very nature of the educational interaction. As W. B. Yeats may have been attributed to have wrote: education is not the filling of pails, but the lighting of a fire. There is every risk, which we have to accept, that the fire may not take, for a variety of reasons. We should not have to resort to desperate measures like dousing the material in petrol and setting things on fire, even though that may ultimately give us what we want.

In my opinion, then, research in makerspaces needs to attend to intrinsic, educative purposes of making. Here, it is probably a good idea to consider again what counts as a makerspace. A certain degree of compromise is necessary here. Overly strict definitions of makerspaces and making activities would unnecessarily limit us to digital fabrication and what may essentially be engineering workshops adapted for novices in an educational context. On the other hand, we should avoid permissive definitions, such as the speaker at a conference I attended who frankly conceded that they were being liberal in definitions of making to include their project in teaching creative writing as a 'making' project, simply because the vagaries of research funding meant that STEM and making/makerspaces were getting all the attention and money. Certainly, the community of self-declared researchers in making have been open and receptive to various forms of making, from digital fabrication [cite], to electronic textiles [cite], craft work [cite], design and technology [cite], and a wide range of other possibilities. Uniting all these approaches is the notion that some form of physical making is involved, such that creative writing is not making even in this broad-based definition. Neither, for that matter is pure programming projects involved in 'making' an application, even though the process shares many similarities to other forms of making. Typical visual arts projects may fall into this spectrum, but for the purposes of this book, I will like to consider forms of making that have, as a direct or unintended consequence, the acquisition of STEM knowledge. Thus, for instance, a kinetic sculpture that requires the artist to grapple with the engineering behind making things behave in a particular desired manner would count as a STEM making project in this definition, while another project that may use items of technology (e.g. light emitting diodes) but only in an accessory sense, would not.

What unites all these diverse processes of making is the physical nature of the interaction. In the process of making, students need to work with materials, to bring matter from one configuration to another, more desired state. Here, this apparently straightforward definition leads us to interesting conclusions if we are careful. For starters, we begin to notice that desirability and intention enters into the consideration. At least at a very basic level of intention, we can see that the maker's intention to change the configuration of material before them is not always something that can be achieved. The tool used may be inappropriate for the task, or the material may not behave as expected. In any case, such an experience would result in a failure, and while we have been largely conditioned in society to only pay attention to successes instead of failure, quite a lot can be learnt from studying failure, such as in the cognitive processes involved in Productive Failure (Kapur, 2008, 2016). As far as I can tell however, Productive Failure methods have not studied the processes in material manipulation, perhaps because of a cognitive bias that may struggle to see cognition in such mundane a task as cutting and joining pieces of matter, as compared to calculating the standard deviation of a set of numbers. Indeed, even the embodied cognition perspective may be dogged by such ideas of a somewhat hierarchical distinction between what has been termed epistemic and pragmatic actions (Kirsh & Maglio, 1994). As proposed by Kirsh and Maglio, epistemic actions are a special kind of action, intended by cognisers as 'physical actions that make mental computation easier, faster, or more reliable—are external actions that an agent performs to change his or her own computational state' (p. 513). On the other hand, pragmatic actions are those actions which 'bring the agent closer to his or her physical goal' (p. 515).

At the outset, this distinction may appear useful. A person physically rotating a Rubik's cube leaves information 'out in the world', sparing the mind from attempting to represent and remember all the face configurations. The same person throwing the cube onto a shelf out of reach seems to be merely achieving a physical goal of putting the cube onto the shelf. Yet, as Loader (2012) explains, this distinction is not as clear as it is initially imagined: we can have no access to the intentions of the person performing the action. Our thrower may wish to glean information on the construction of the shelf, and the sound the cube makes as it lands on the shelf could provide valuable information. More significantly, 'throwing a cube to a shelf' may be composed of a series of actions, such as physically moving closer, changing one's perspective to better gauge distance, each one of which contributes information to the macro process of throwing, and should be rightfully considered as an epistemic action. More generally, and to the point of this book, is the notion that we should stop interpreting actions as merely pragmatic and somehow lower in hierarchy than a cognitive operation.

The mind-hands dichotomy has been an pernicious influence in the way we have been thinking about the relative valuing of the different types of learning, and this has probably been made worse by the industrial revolution and the subsequent attempts on the part of capital to lower labour costs by Taylorist time and motion studies (Carr, 2014, p. 107). Such studies sought to 'scientifically' decompose the complicated steps artisans took to create something, into a series of steps, each so simple that any 'moron' could perform it (Waring, 1991). In the process, this reduced the process of *making/crafting* into a process of *manufacturing*, lowering the cost of labour and collectively putting the job security of the bulk of the organisation's employees at extreme risk. As retold by Carr, this was the genius of Henry Ford and his production line, in reinventing the role of the employee. The fact that much of the work on production lines of all forms are being threatened by robotics and generalised automation today should not be seen as a victory for robotics engineers, but rather as evidence of the dismal conditions workers on production lines have had in the intervening period between the invention of the line and contemporary robotics. Having personally worked one single day as a human robot filling paint cans in a factory, I can attest to the utterly soul draining and dehumanising nature of such work, and how little of the work required the working of an operational mind. It had to be, as the factory neglected to provide any form of breathing protection from the volatile solvent fumes; I refused to return to work on health grounds and did not even bother to collect my one day's worth of income.

More distressingly, Carr suggests that we should not think that this process of separating thinking and doing is only limited to manual labour. As anyone who has worked in, or had to interact with, general office administration can attest to, the production line mentality is thriving in 'white collar' labour too. How else do we explain the compartmentalised handling of routine administration and the frustration we encounter with the attitude of 'it's not our department's responsibility'? Carr's critique also suggests that any routinisable form of labour will eventually find its way to being automated by newer and ever more effective technologies, from piloting aircraft, medical diagnosis, and legal discovery. Even creative operations may be at stake when we consider such advances as parametric architecture as a means to develop new designs, by simply specifying parameters (locations of windows, doors, maximal dimensions; location of utility conduits, etc.) and having a computer program figure out how to 'connect the dots' around these constraints. The larger point here is the increasing necessity for us to think deeply about the nature of humanity, the role we expect our students to play in uncertain futures, and the social justice aspect of our educator's role in gatekeeping people into decision-making and decision executing roles.

Considering the fact that we are building increasingly sophisticated computer systems that can make rapid manipulations on representations of our knowledge, there may be a progressively weaker case for *solely* privileging competence in abstract manipulation. This is not necessarily to say that we should encourage our students to professions in the trades; nonetheless, such a recommendation may not be entirely wrong. Given the propensity for complicated mechanisms to break down in unpredictable ways, such forms of work are likely to not be routinisable and replaceable with automation. Consider Crawford's (2009) motorcycle that would not start: a philosopher-mechanic[2] like Crawford who specialises in old motorcycles derives great satisfaction and cognitive labour from the process of debugging because there are no sophisticated electronic systems such as On Board Diagnostic (OBD) computers. On machines with OBDs, the mechanic's role is reduced to simply reading the error code, consulting a manual for a translation, and then swapping out the appropriate module—again, another incidence of humans playing a roboticised role in the system of machine maintenance. In contrast, Crawford working on his old motorcycle retains much human agency as an investigator who studies the symptoms, makes hypotheses about what could be wrong, and then progressively tests each hypothesis until the root cause is found. Such a practice, it must be said, is also the case for the physician attending to a 'broken down' human—and these usually come without an OBD; I will have more to talk about the process of abductive reasoning that unites all these processes. It is sufficient however to note Carr's troubling message that 'expert systems' have been attempting to routinise healthcare (among other fields) by automating diagnosis. In stark terms, the physician in many cases has been reduced to human user interface (an 'OBD port') for a computer-based system that is supposed to reduce misdiagnosis and physician error in prescription.

Given that the institution of schooling serves in large part to prepare students for future economic participation; or more generally, school prepares students for living, a large part of life for most will inevitably be economic activity. At the same time, it is fairly well accepted that schools ought to serve both conservative and progressive social purposes: students need access to multi-generational projects that are larger than themselves, yet we must not be satisfied if students only ever rehearse old knowledge claims and never develop on a trajectory of owning and developing new ideas. While no attempt at predicting the future is ever certain, there is a certain sense that the long-term trajectory of increasing automation and increasing capability of computers will not see any abatement in the near term. It seems to me that our response as educators to this (and other) impending future(s) may be one of two possibilities: we may decide to shift our educational attention towards preparing students for a topical attention to these ideas in a wholehearted embrace, or we may decide that a more critical, more careful approach is more desirable. As will become more apparent as the book develops, I advocate the latter, more critical approach, if only because I believe that the essence of an educative experience is its opening for students a greater latitude of possible action. Making in makerspaces with contemporary technological implements *can* have educational uses, but no assumption is made that simply because a technology is new and perhaps even 'proven' by research to be effective, that students *ought* to use it. The problem here is one of ends, and not only the means.

Making, not manufacturing

To summarise this chapter, and to come to some implications for the venture of makerspaces for science education for innovativeness, the main proposition here is that we should be concerned with school in general, and makerspaces in particular, as sites for making, and not manufacturing. I arrive at this conclusion from considering the nature of 'working with one's hands' in the process of making things. The point of makerspaces should not be necessarily and primarily as a site where one manufactures understanding of sophisticated representations, as quite some ongoing research is demonstrating. The goal here is to establish an intrinsic rationale for making as an educative activity. To do this, I have reviewed the concept of embodied cognition as a means to discern 'what is going on' when people are engaged in the act of manually manipulating materials. The embodied cognition perspective allows us to consider that the 'minds-hands' dichotomy as false, or at least unhelpful; further, the privileging of thought over action as unnecessary. It is not as if people involved in making things, especially in non-routine situations, develop plans 'in the head', fully formed, before implementing these actions on the material world. The suggestion here is that such a perspective has been largely the result of a very successful industrial mode of thinking that separates the planning and cognition from the implementing. While I will more formally engage with this notion of industrial thinking in a later chapter, it is sufficient to say here that this perspective is not helpful to this book's project. Given the somewhat inevitable nature of increasing computing and technological prowess, there may be an imperative to think clearly about continuing to make space for humans to continue being human and not serve as merely moist appendages to programmed machine logic. In places of work, we can read critiques such as Crawford's meditations on the dehumanising nature of contemporary work, and Carr's warnings on how contemporary technologies are removing agency and autonomy, to infer some degree of truth into the popular aphorism that in the imminent future, jobs will be of two forms: one where humans tell computers what to do and another where computers tell humans what to do.

To be clear, this is not intended as a neo-Luddite argument for a return to some mythical past; I accept that Pandora's box has opened and there is simply no return. Yet, there is a sense that the metaphor of Pandora's evils is not completely accurate—while the evils are almost literally forces of nature, technologies by definition are products of artifice. Products of artifice are essentially statements of intent made by the people who design them and mark an arbitrary deviation from the underlying science of understanding how nature works. In other words, there is nothing inherent, essential, or necessary in the design of technological objects. For example, our smartphones can track our location and spy on our online and real-world activities, but that is not a consequence or a requirement of the underlying technology; some organisation of people decided it was important to do so and managed to convince us that it was necessary. As educators, it becomes increasingly important that we educate our students to become more critically aware of the distinction between the necessary and the arbitrary (rehearsing ideas from the previous chapter). In this case, learning how to use technology in makerspaces must be seen only as the first step in their education. In this first step, it is important that students reconnect with the way things are in themselves, rather than get caught up in arbitrary representations that may limit the available information one may acquire from interacting with the things themselves. Continuing the smartphone example, it may be valuable to conduct lessons on the schematics and operating principles behind the surveillance technologies. However, the educative value of practical activity of making extends beyond 'applying' the abstractions onto real-life objects. Making activities here ought to extend opportunities for students to break things (metaphorically and literally), so as to find the limits and caveats of the representations; to develop a feel for the connections between the phenomena and representations such as to understand the arbitrary nature of the connection.

In this regard, the exhortation for educators is to privilege activities where things are *made*; educators, often with the best intentions of leading a class of students through the efficient conduct of learning activity, may organise *manufactured* experiences instead. Such experiences can be the result of traditional science laboratory investigations, typically consisting of teacher-led demonstrations and recipe-driven activities where particular conclusions are expected. To be fair, the recent emphasis on inquiry processes in scientific investigations [cite] goes a long way towards attending to this issue, but the overwhelming desire for schools adopting these practices is still the acquisition of high-value theoretical abstractions. Less prevalent are attempts for students to acquire or develop inventiveness and to appreciate the gap between representation and phenomena.

How to makerspace?

If the above provides a rationale and vision for what making and makerspaces can do for students, there still remains the question of how such a makerspace can be run, especially within a formal schooling context where accountability demands can steer teachers away from much else besides preparing students for standardised examinations. In the following, I will discuss cases from my research in schools in Singapore. These cases are the result of studies into setting up and characterising makerspaces across four projects that I have been involved with. In order to preserve the participants' anonymity, all names used will be pseudonyms. While these studies have had deliberate purposes within the research grants that funded their study, I will report selected aspects of these makerspaces, as case studies. Methodologically, I derive support from researchers such as Flyvbjerg (2006), Ruddin (2006), and Yin (2009), who provide strong rationales for the rehabilitation of case studies. Case studies, these authors claim, are not to be treated as glorified anecdotes, hypothesis generation devices, or ungeneralisable pilot studies for later investigations. The goal of reporting these cases is to demonstrate a form of existence proof of the possibility for particular visions of making and makerspaces. Educational interactions are rich phenomena that take place across the entire spectrum of possible analysis modes. While it can be routine to engage in virtually scientist research, reducing educational phenomena to psychology, and then further into cognitive psychology, my belief here is that such reduction can be ill-advised as it can be close to impossible to separate contextual factors from these forms of 'microscopic' analyses. Even if such a position was possible, the question of the desirability of the particular forms of interactions cannot be answered by merely demonstrating the psychological efficacy of an intervention. For instance, prior research on behaviour modification approaches to learning may have produced good science which showed its effectiveness. Yet, not many educators today will wish to be associated with behaviour modification because it is perceived as removing learners' autonomy from the interaction. In any case, educational interactions, especially those with high ecological validity, are likely to be ephemeral collections of coincidences; the goal of education research should not necessarily be the generation of law-like description of decontextualised patterns that are portable to other contexts, but the generation of insights that are intelligently adapted by agentic individuals. There is unlikely to be a pill to treat education, nothing that can be abstracted from one context to treat another educational context. Such is at least the conclusion of Allan Luke (2011), prominent scholar and researcher of school systems at the American Educational Research Association distinguished lecture: the educational success of places like Finland and Singapore cannot be reduced to singular measures on international comparative assessments, and seeking the cause of such successes only in classroom performances may be misleading. With that in mind, the cases presented in this book serve as meaningful descriptions of the tensions and often contradictions that may occur, besides the potential for bearing insightful approaches to the problem of implementing makerspaces with a particular vision of its educative potential.

The concern in this chapter has been with the relationship between the tacit knowledge of the phenomena and its more formal representations. If we believe that such tacit knowledge is meaningful to students' learning, an interesting question is how schools can develop meaningful ways to recognise that tacit knowledge is being developed. One possibility comes from the model I observed at Able High School. With students from grades seven through twelve, Able is a school which had attracted many talented students of the natural sciences and mathematics, and the school administration had decided to work on this strength to become a centre of excellence for science and mathematics instruction. This meant that they managed to secure permission from the Ministry of Education, which oversees the funding and administration of public schools in Singapore, for somewhat greater autonomy in order to run their custom programmes and assessments. As many of the school leadership have a passionate concern for the natural sciences, being graduates of science and mathematics disciplines, it was easy for them to notice, before the contemporary popularisation of the maker movement, that many of their students had theoretical knowledge, but inadequate appreciation of the experimental procedures surrounding the production of knowledge. In this regard, Alfred, one of the heads of department in Able, quickly took advantage of a school structure to address this.

The school had, as part of its graduation requirements, a capstone student project for students at the end of their grade eleven studies. This project was modelled after the graduation requirement for undergraduate students and was research based in nature, getting students to practice creating new knowledge.

As the school could afford to be ambitious with its students, it set high standards for this project, matching up students with mentors from the local universities and other research organisations. These student projects ended up of rather high quality, often meeting or occasionally even surpassing the typical standards for university undergraduates. The school invested in research grade laboratory equipment, which reduced students' need to head to the university laboratory all the time. This solved some of the logistical problems, but introduced some of its own. These machines were not 'plug-and-play' appliances. While the operating instructions were ostensibly straightforward, it was not like using a refrigerator: open door, store things, close door. Getting a good result on a spectrometer required that samples be prepared adequately, that the optical path was not obstructed, that calibration against known standards were conducted frequently, and so on. Soon enough, student projects required sufficient material manipulation and the creation of customised experiment rigs that a general-purpose workshop was needed to support these projects. As the number of cases grew, teachers at Able started to notice that some of these experimental set-ups were, in and of themselves, intricate assemblies that demonstrated a high degree of thought, planning, and understanding of the scientific background. This need to support a general-purpose workshop grew to become a makerspace, whose general organisation resembled a student-led 'homework club' of students working on their capstone projects. This homework club soon expanded to include junior year students, who found an affinity to tinkering and making things, as part of their scientific investigations, or out of plain curiosity. Many a time, students would come across interesting videos online of amazing phenomena, and simply wanted to copy what they found. Often, however, these students moved on from copying, to trying to figure out the science behind the phenomena, and then into creating new knowledge for themselves and others. When it came to trying to analyse and improve on their experimental set-ups, a certain degree of precision, repeatability, and automation was often required. For this, students then found equipment like laser cutters, 3D printers, and microcontrollers very useful.

For instance, Beverly, a grade 11 student at the time I studied her work, got interested in the ability of solenoids to propel steel projectiles. While the basic scientific explanation of the phenomena was well understood by students her age, she was not content to merely watch a nail jump; instead, she wanted to develop a powerful enough 'coil gun' to actually do some real damage. While many other schools would have quickly shut down her ambitions on the grounds of a concern for safety and a lack of insight into how such an activity could be educational, teachers at Able were quick to recognise the potential for in-depth learning that this situation posed. Beverly soon found that she needed to increase the energy transfer from electric to kinetic and figured out her approach was to begin with an excess of electrical energy, and to dump all this electrical energy in one brief moment. This meant that she had to make use of high capacitance capacitors and a switching circuit that would not destroy itself when the resulting huge current passed through it. The initial design worked well, but did not deliver high enough projectile velocities, so she decided to add subsidiary coils to further increase the energy delivered. Immediately, this led to synchronisation issues which were not trivial to solve: a moving projectile through a coil could actually be repelled away from entering, or attracted back into the coil, if the current pulse was mistimed. Alfred, who was the head of department responsible for school partnerships, gave the school's blessings for her to work with engineers at the local Defence Science research organisation, where she developed her interest in electrically powered kinetic propulsion systems to a very high level. She started work on her capstone project characterising electromagnetic propulsion devices and in her spare time worked on her multi-stage coil gun, which eventually needed high current capacity transistors, and a microcontroller to handle the timing requirements.

This case presents us with demonstration of some of the principles elaborated upon in this chapter. In traditional approaches to school-as-learning-factory, with an effort to standardise outputs and control production processes, the most likely process is one where the theoretical foundations for electromagnetic propulsion are introduced *en masse* in lecture format, before an aptitude test of some sort is administered, and individuals considered worthy are advanced to the next step where they can now 'apply' their abstract conceptual knowledge in the practical context of creating such a device. Instead, Able Secondary has flipped this model on its head, starting off with the subjective and qualitative experiencing of interesting phenomena with which the students had been empowered to follow wherever their curiosity led them. Beverly's purpose was relatively simple: build an electromagnetic device which could accelerate a projectile to a high enough velocity. Yet, this simple ambition led her down a deep 'rabbit hole' of sorts, to very sophisticated theoretical understanding of high current switching, microcontroller interfacing, programming, measuring projectile velocities with high-speed cameras, not to mention all the practical skills required to fabricate a small circuit board, solder components together, and assembly of the entire device with enough mechanical precision such that the acceleration stages were in the right place given the timing constraints. In the end, she created a device that, she was proud to tell me, could accelerate an approximately 3-cm-long nail and cause it to penetrate both walls of an aluminium drink can.

To be clear: the instructional design principle for makerspaces and STEM education illustrated by this case is the importance of giving students the opportunity to pursue their own ambitions. Relatively simple ideas ('I want to make this go faster') can lead to high degrees of sophistication; there is no contradiction here that simple goals can be difficult to achieve and require an extensive amount of science, mathematics, technology, and engineering. The purpose of the practical experience in makerspace is to create the intellectual demand, and the role of the instructor is to educate the desire, in the sense of letting the student know just what it is that they are supposed to want. The scientific point of view desires a particular order and structure to the way phenomena is organised and understood; these perspectives are often not intuitive and are tacit in that they only make sense in relation to the subjective experiencing of phenomena.

Summary

The anti-intellectual approach to knowledge and learning can be counter-intuitive, especially for educators with a mechanistic desire for education to be controllable and guarantee a certain minimum basic measurable result. Our common, everyday experience with manufacturing things in a repetitive manner can lull us into a sense that humans can and should be 'developed' using a similar approach. Unfortunately, humans do not learn in this manner; we are not computers or abstract symbol processors downloading data and instructions into our brains in order to perform calculations on them. We require grounded experiences, minimally as starting points to make metaphorical extensions from, or as means for us to acquire the tacit knowledge necessary to grasp the aspects of the experience that we cannot represent. The process of education, unlike the manufacture of industrial goods, requires the active, agentic cooperation of those being educated (or else the activity should rightly be called indoctrination), and activities in makerspaces should be seen as an effort to educate the desire of individuals. What makes the practice of scientists, mathematicians, engineers, technologists distinct from others is in the way that they choose to focus on particular aspects of the phenomena, and how they desire particular outcomes that make sense within their particular framing. The development of novices to 'see' these points of view can only be considered complete if they can use these lenses independently of instruction to do so. In consideration of all these, it is therefore useful to take away a simple dictum for makerspace instruction: we should aim to *make*, to *invent*, and not manufacture.

Notes

1 Here, 'economistic' is used in a similar manner to the slur of 'scientistic' goals—as only having the external form, but ultimately lacking the substance.
2 Crawford actually has a Ph.D. in philosophy and was employed in a Washington DC think tank for some time before he quit in disgust at the kind of work he was asked to do.

References

Abrahamson, D., & Kapur, M. (2018). Reinventing discovery learning: A field-wide research program. *Instructional Science, 46*(1), 1–10.

Alibali, M. W., & Nathan, M. J. (2012). Embodiment in mathematics teaching and learning: Evidence from learners' and teachers' gestures. *Journal of the Learning Sciences, 21*(2), 247–286.

Barsalou, L. W. (2008). Grounded cognition. *Annual Review of Psychology, 59*, 617–645.

Barsalou, L. W. (2010). Grounded cognition: Past, present, and future. *Topics in Cognitive Science, 2*(4), 716–724.

Barsalou, L. W. (2016). On staying grounded and avoiding quixotic dead ends. *Psychonomic Bulletin & Review, 23*(4), 1122–1142.

Bergen, B., & Feldman, J. (2008). Embodied concept learning. In P. Calvo & T. Gomila (Eds.), *Handbook of cognitive science: An embodied approach* (pp. 313–332). San Diego, CA: Elsevier.

Biesta, G. (2016). *The beautiful risk of education.* Abingdon: Routledge.

Carr, N. (2014). *The glass cage: How our computers are changing us.* New York, NY: W. W. Norton & Company.

Clark, A. (2003). *Natural born cyborgs.* Cary, NC: Oxford University Press.

Clark, A., & Chalmers, D. J. (1998). The extended mind. *Analysis, 58*(1), 7–19.

Clement, J. (1989). Learning via model construction and criticism. In J. A. Glover, R. R. Ronning, & C. R. Reynolds (Eds.), *Handbook of creativity* (pp. 341–381). Boston, MA: Springer US.

Collins, H. (2010). *Tacit and explicit knowledge.* Chicago, IL: University of Chicago Press.

Crawford, M. B. (2009). *Shop class as soulcraft: An inquiry into the value of work.* New York, NY: Penguin Press.

DeBoer, G. E. (1991). *A history of ideas in science education: Implications for practice.* New York, NY: Teachers' College Press.

Dennett, D. (1991). *Consciousness explained.* New York, NY: Back Bay Books.

Dennett, D. (1998). *Brainchildren.* Cambridge, MA: The MIT Press.

Dennett, D. (2009). Darwin's "strange inversion of reasoning." *Proceedings of the National Academy of Sciences, 106,* 10061–10065.

Flyvbjerg, B. (2006). Five misunderstandings about case-study research. *Qualitative Inquiry: QI, 12*(2), 219–245.

Fodor, J. (1975). *The language of though.* New York, NY: Thomas Crowell.

Fyfe, E. R., McNeil, N. M., Son, J. Y., & Goldstone, R. L. (2014). Concreteness fading in mathematics and science instruction: A systematic review. *Educational Psychology Review, 26*(1), 9–25.

Horst, S. (n.d.). The Stanford Encyclopaedia of Philosophy (Spring 2011). Retrieved from http://plato.stanford.edu/archives/spr2011/entries/computational-mind/.

Kapur, M. (2008). Productive failure. *Cognition and Instruction, 26*(3), 379–424.

Kapur, M. (2016). Examining productive failure, productive success, unproductive failure, and unproductive success in learning. *Educational Psychologist, 51*(2), 289–299.

Kirschner, P. A., Sweller, J., & Clark, R. E. (2006). Why minimal guidance during instruction does not work: An analysis of the failure of constructivist, discovery, problem-based, experiential, and inquiry-based teaching. *Educational Psychologist, 41*(2), 75–86.

Kirsh, D., & Maglio, P. (1994). On distinguishing epistemic from pragmatic action. *Cognitive Science, 18,* 513–549.

Lakoff, G., & Johnson, M. (1999). *Philosophy in the flesh.* New York, NY: Basic Books.

Lakoff, G., & Núñez, R. E. (2000). *Where mathematics comes from.* New York, NY: Basic Books.

Lawson, A. E. (2005). What is the role of induction and deduction in reasoning and scientific inquiry? *Journal of Research in Science Teaching, 42*(6), 716–740.

Lawson, A. E. (2010). Basic inferences of scientific reasoning, argumentation, and discovery. *Science Education, 94*(2), 336–364.

Loader, P. (2012). The epistemic/pragmatic dichotomy. *Philosophical Explorations: An International Journal for the Philosophy of Mind and Action, 15*(2), 219–232.

Luke, A. (2011). Generalizing across borders: Policy and the limits of educational science. *Educational Researcher, 40*(8), 367–377.

Martín-Páez, T., Aguilera, D., Perales-Palacios, F. J., & Vílchez-González, J. M. (2019). What are we talking about when we talk about STEM education? A review of literature. *Science Education, 10,* 165.

Mayer, R. E. (2004). Should there be a three-strikes rule against pure discovery learning? *The American Psychologist, 59*(1), 14–19.

Nathan, M. J. (2012). Rethinking formalisms in formal education. *Educational Psychologist, 47*(2), 125–148.

Nathan, M. J., Walkington, C., Boncoddo, R., Pier, E., Williams, C. C., & Alibali, M. W. (2014). Actions speak louder with words: The roles of action and pedagogical language for grounding mathematical proof. *Learning and Instruction, 33,* 182–193.

Nemirovsky, R., Rasmussen, C., Sweeney, G., & Wawro, M. (2012). When the classroom floor becomes the complex plane: Addition and multiplication as ways of bodily navigation. *Journal of the Learning Sciences, 21*(2), 287–323.

Pinker, S. (1997). *How the mind works.* New York, NY: Penguin Books.

Pinker, S. (2002). *The blank slate.* New York, NY: Penguin Books.

Polanyi, M. (2009). *The tacit dimension.* Chicago, IL: University of Chicago Press. (Original work published 1966.)

Rowlands, M. (2010). *New science of the mind: From extended mind to embodied phenomenology.* Cambridge, MA: MIT Press.

Ruddin, L. P. (2006). You can generalize stupid! Social scientists, Bent Flyvbjerg, and case study methodology. *Qualitative Inquiry: QI, 12*(4), 797–812.

Searle, J. R. (1980). Minds, brains, and programs. *The Behavioral and Brain Sciences, 3*(3), 417–424.

Selwyn, N. (2016). Minding our language: Why education and technology is full of bullshit … and what might be done about it. *Learning, Media and Technology, 41*(3), 437–443.

Waring, S. P. (1991). *Taylorism transformed: Scientific management theory since 1945.* Chapel Hill, NC: UNC Press.

Yin, R. K. (2009). *Case study research: Design and methods* (4th ed.). Thousand Oaks, CA: Sage Inc.

4
DESIGN AS A PROBLEM FOR SCHOOL THAT REVEALS THE PROBLEM OF SCHOOL

When we think about the future and how schools might prepare our students for it, it has become a prevalent notion in the last decade or so to consider the design method as the method of choice. Easily displacing 'the scientific method' as the key method for our times, the design method is lauded to be behind the success of large multinational corporations, especially those involved in the 'high tech' industry. Most notably, the spectacular success of Apple Corporation has been credited to the use of design; not merely in the aesthetic sense of improving the way things look, although former CEO Steve Jobs' description of 'lickable' user interfaces was an excellent distraction. The genius of the design method is exemplified in the manner in which people interacted with their products. Even from the moment the shipping box is received, there is no 'friction' in the user interaction: the boxes come with easy open seals that manage to protect its contents securely while still ensuring users need not require any tools whatsoever to get at the products. Computers used to be shipped with thick user manuals, and now we have tablets which young children take to intuitively, and we fall into the illusion that we have a 'net generation' who are somehow wired differently than previous generations. Behind this facade of frictionless-ness lies a great deal of work; from the diverse disciplines of anthropology, psychology, philosophy, cultural studies, and not to mention, of course, the science, technology, engineering, and mathematics (STEM) knowledge bases which are most often directly credited with the success of these sorts of enterprises. Hailed as a method that better guarantees economic success, the design method has been taken up as the *de facto* method for organising makerspace activity. Yet, we already begin to see the outline of challenges to conventional schooling: for a start, most schools are organised around a reductionist, discipline centric system, while the successful design requires an adroit balance and selective drawing from the diverse insights that each 'mode of seeing' affords. Design is an archetypal creative activity, whose

DOI: 10.4324/9781351116220-4

outcomes are judged as ideal if a certain degree of transgression occurs; schools are usually more concerned with the effective reproduction of knowledge, dominant perspectives of schooling leading to such contradictions as the assessment requirement of being able to discriminate designs to meaningless differences on some arbitrary rating scale. Last but not least, many schools have been caught up in technocratic modes of being, focussing on achievable, definable, close-ended goals rather than an open-ended pursuit of tenuous goals such as 'excellence', 'elegance', or 'wisdom', ideals that designers (ought to) pursue.

In this chapter, I want to take up the challenge of considering design as a school subject and specifically as an instructional method for makerspaces. As with previous chapters, I intend to consider deeply the knowledge of design and draw some implications to curriculum and pedagogy. The knowledge basis for design are fairly new, given design's entry into the academy, and thus there has not been a sufficient movement to consider the 'nature of design' for design (and pertinent to this book, STEM) education, in the same way that implications from research into the 'nature of science' have informed science education in the past 30–40 years. Instead of thinking about design as a fundamentally mysterious activity performed by inspired genius, I want to think about design in its mundane form, considering design not in the glamourous settings of professional design studios, but in day-to-day usages of design knowledge for all sorts of purposes. Instead of thinking about design as the precursor to industrialised manufacturing process, I seek instead to consider how individuals may use design and interact with materials in order to *make* artefacts that express the particular intention of the designer. What role does the social play in the process of design, and how do designs interact with the social, are questions that will have implications for how design, in makerspaces, is implemented. Even more ambitiously, I wish to argue that these ideas provide us with a new perspective with which we can, and should, reconsider how schools are organised.

What is design?

In this age of machines that seem to work effortlessly, reaping investors in Silicon Valley corporations vast sums of money, it may seem that the model of 'Designed in California, Made in China' may be the model to emulate. Indeed, much of the current attention in education appears to be directed towards the recent trend of 'STEM'. With rather vague definitions, the amorphous quality of STEM has allowed implementors a rather wide range of activities which still fall under the banner of STEM. It should not be surprising to learn that, for instance, at a recent international education research conference, a presenter declared her creative writing workshop as a 'maker' project, even unabashedly declaring she did so because that was where the research funding was. Much of STEM and makerspace activity appears to be organised around engineering design methods, with problem finding and creative problem solving as focal activity (e.g., Capraro, Capraro, & Morgan, 2013). The proliferation and relative success of what may be called the 'Californian ideology' (Barbrook & Cameron, 1996) of design has led to works by authors such as the Kelley and Kelley (2013), whose stated aim is to recover in individuals their (lost) confidence in their own creativity. According to them, the prevalent mode of thinking taught in schools has been one of reductionist analysis, leading to intelligent individuals whom in their own experience as design consultants are excellent executors of business strategy, but self identify as 'uncreative'. Design, they continue, may be understood as a collection of discrete tactics such as, among other things: (i) anthropological tools (they do not refer to it as such) directed towards understanding the human needs and problems that people encounter; (ii) project management techniques privileging action over cogitation; and (iii) team tactics that create localised cultures that nurture behaviours likely to result in creative outcomes. While much of the book's audience is the busy business executive seeking 'five step methods' to improve innovativeness in the workplace, at least three valuable insights from the book are useful for our discussion here about organising educational contexts for design instruction. Kelley and Kelley (*ibid.*) note that good design stands at the confluence of technical feasibility, social desirability, and business viability; that the designerly intention is an almost magical quality that deserves nurturance; and failure is an inherent part of the design process and needs to be embraced as such. This leads to such pithy Silicon Valley sayings as 'fail often so that you can succeed earlier' or the colourful imagery of founding a start-up as flinging oneself off a cliff and hoping to assemble a working aircraft before one meets the more likely end of an untimely collision with the hard reality of the earth (García Martinez, 2016).

Put in more general terms, the observations of such successful designers point towards the insight that the act of design is a motivated, intentional synthesis of a wide base of knowledge in order to identify and then subsequently solve a problem. Candidate solutions, as works of creativity, are almost always guaranteed to fail in their first few iterations for a variety of factors ranging from a misperception of the problem to a misapprehension of the technical feasibility of the project. Already, we begin to see the outlines of a problem for schooling: for almost all schooling contexts emulating the industrialised, western, English speaking system, schools function as gatekeeper and introduction to the logic of industrialisation. Built around the principle of reductionism, schools legitimised the idea of the existence of diverse, discrete disciplines, instead of a holistic experiencing of phenomena. In effect, schools maintain the relative standing and academic privileging of some forms of knowledge over others, resulting in situations where some forms of knowledge are deemed useless, especially when viewed in economistic terms (Midgley, 1990). As for schools being gatekeepers, mistake making and, especially, the idea of academic failure can weigh heavily on students' willingness to take risks even when explicitly encouraged as part of the intended learning outcomes for a class on design (e.g., Tan, Lee, & Ng, 2017). Nurturing creative intentions can also stand in contradiction to typical schooling conditions which often expect compliance with established norms.

The purpose of school is more often the bringing of students into convergence with well-defined curriculum standards; encouraging students to diverge often upsets teachers who can find it difficult to manage standardised assessment.

To return to design, we note that design as an academic subject is perhaps one of the youngest 'disciplines'; design as an independent activity that could be studied simply did not exist prior to the industrial revolution, as design and making were co-extensive activities. It was the industrial revolution that introduced the notion that products could be manufactured *en masse* by reproducing a basic form, that the possibility of a separation of planning and making became important. As a result of this relative youth, and the weak boundary maintenance between design and other disciplines, design has compared poorly in comparison to such established and well-defined disciplines such as the natural sciences and mathematics. Because of its relationship with industrial manufacture, design has the unfortunate association with the aesthetic beautification of items for the enhancement of the products' appeal to potential consumers. To be sure, design is interested in aesthetic enhancement, but is involved in a lot more than that. Every well-designed consumer electronics product (for instance) appeals to the aesthetic senses not only from its external appearance, but also as alluded to above, from the effortless functioning of its internal parts that are designed to serve the particular intended function. Engineers are heavily involved in the process of design, as are architects, social planners, game designers, and of course the numerous forms involved in modifying the appearance of items such as clothing, furniture, jewellery, and so on.

While much of design appears to draw from, and can be considered a process of, art, it may be useful to consider Herbert Simon's influential characterisation of design as a science of the artificial (Simon, 1968/1996). According to Simon, design exists at the interface between the inner and outer worlds of the designed artefact; its concern is with 'attaining goals by adapting the former [the inner environment] to the latter [the outer]' (p. 113). Tantalisingly, he suggests that a 'science' of design is possible, consisting of 'a body of intellectually tough, analytic, partly formalizable, partly empirical, teachable doctrine about the design process' (p. 113). Because Simon's 'home discipline' was that of artificial intelligence, and the book held the ambition to create a science of design, given its context of mass academic 'physics envy' in the post war years, one of Simon's solutions was to reduce design to a procedure of satisfying constraints placed upon a designed system via a version of linear programming. Given his claim of design as a process of making the 'inner environment' conform to constraints placed upon the 'outer environment', he describes a mathematical process of placing weights upon decision-making factors and then attempting to compute the maximum or minimum value of the outcome as desired, but only as a starting point. Acknowledging the existence of a class of computational problems which are essentially intractable, requiring too much time, he further proposes that design solutions are often 'satisficed', a portmanteau of 'satisfy' and suffice: solutions are generated and tested against criteria until a sufficient solution that satisfies the problem constraints is found within the time allocated for the search. Part of this process must be the various heuristics that designers utilise to produce possible solution candidates. Through experience, insight, and intuition, expert designers are able to quickly rule out particular candidate classes of solutions as not productive, and either prune off or add branches according to design constraints or intentions as needed.

Is design an art or a science?

Perhaps most significant to this book is the general question of whether design is an art or a science. This question is useful to ask because, at least at the outset, it may appear that scientific knowledge may be expressible as a set of abstractions that can be communicated absent a context. If this were the case, the possibility arises that design can be taught *en masse*, in lecture form. If, on the other hand, design is an art, then it is more likely the case that an apprenticeship model is needed. Besides the pedagogical concern of how best to communicate knowledge in design, there is also the question of academic prestige. To have a scientific, objective basis for one's work was (and to a great extent still is) perceived to be a marker of prestige, as compared to a more aesthetic, subjective point of view. As alluded to above, and expanded upon in an earlier chapter, a significant amount of intellectual work in the 20th century has been directed to explaining the spectacular success of the sciences in explaining, predicting, and controlling the natural world. For design, its relatively recent entry into the university as a discipline has been marked by the debate among its exponents attempting to associate it, as with Herbert Simon had, with the natural sciences, or as others have, with the arts. Even less than 15 years ago, Nigel Cross (2006) has surmised that not enough has been said about the ontology and epistemology of design. There are pronouncements such as Simon's, but much of these are not well supported by actual studies of designers as they got about their work.

As summarised by Cross (2001), the 1920s were a period where there were attempts to 'scientise' design; indicative of the era of positivistic overconfidence, Cross recounts Dutch artist and architect Theo van Doesburg as remarking that 'Our epoch is hostile to every subjective speculation in art, science, technology, etc. The new spirit, which already governs almost all modern life, is opposed to animal spontaneity, to nature's domination, to artistic flummery. In order to construct a new object we need a method, that is to say, an objective system'. In the 1960s, and in the wake of a science envy after the then-recent events of the World War which seemingly ended with the use of nuclear weapons, attempts were made again to define design, as Herbert Simon did, as a science of the artificial (1968/1996). About a decade later, there was an almost inevitable pushback, by designers who rejected the 'machine language, the behaviourism, the continual attempt to fix the whole of life into a logical framework' (Jones, 1997, in Cross, 2001, p. 50). This period was also the time where designers appeared to have worked out the main distinction between the sciences and design as the different kinds of problems that each faced and the solutions that each generated. Generally, the consensus was reached among design researchers that science pursued close-ended explanations of existing systems, resulting in general principles that could apply to many contexts, while designers are interested in open-ended, ill-defined problems, creating context-bound solutions that did not previously exist. In the 1970s, the dominant epistemology of science was also in flux, troubled as it was by the publication of Thomas Kuhn's *Structure of Scientific Revolutions* (1962/1996), Paul Feyerabend's *Against Method* (1975/1988), and Bruno Latour's *Laboratory Life* (1979/1986), among others. The towering edifice of objective, positivistic science was under attack, or at least being closely examined, and it was easy to resist the temptation of scientism. Scholars especially in the humanities were starting to assert the unique value of their discipline, as researchers began to use anthropological methods in the places where science was practiced and found that the way in which science was actually practiced did not quite match up with how science was reported to be practiced. More of this will come in a subsequent chapter.

While the exact consequences of the critique are not the point here, it remains that design studies began to stand on its own in the 1970s, apart from considerations of whether it was a science or not. 'The' design method began to be formalised and taught in schools and universities; while 'the' scientific method was perceived to be the method for obtaining certain truth about the universe, design method was to be the method for the creation of artefacts. This method, as overly simplified in the sciences as in design, caricatured the manner in which, after the fact, practitioners reported on how they proceeded (hence the scare quotes). Design, it is claimed, proceeds in a cyclical or non-linear manner through phases including: (i) user needs analysis; (ii) problem identification; (iii) brainstorming and ideation; and (iv) prototyping and testing. While there is some degree of simplification, this does not mean that methods do not exist. Just as the scientific method exists as a useful heuristic for first timers learning its practice, the design method can be useful as a means to induct learners. For now, not much more needs to be said about the design method except to note that, as I introduced earlier, the predominant discipline that underwrites the 'user needs analysis' portion of the design method is fundamentally anthropological in nature; problem identification is informed by philosophical studies that tell us what are good problems we ought to solve; brainstorming and ideation has been studied extensively by psychological methods since Guilford's (1950) exhortation to the American psychological academy to understand creativity; prototyping and testing are the domain of scientific analysis. In other words, design is a collection of a family of methods oriented towards the creation of *artefact*, be it some configuration of materials, or a social organisation of people, or a particular method of operation, among the various concrete and abstract forms that can be arteficed.

Nigel Cross (2001) provides some clarity about the relationship between science and design, by considering that there are potentially three major movements or interpretations of the relationship: (i) scientific design; (ii) design science; and (iii) a science of design. Scientific design arose in response to 'scientific methods' in an attempt to distinguish the modern, industrial design, especially of industrial era (and later artefacts) as distinct from an intuitive design that marked the craft-oriented pre-industrial age. Contemporary design processes are dependent on scientific knowledge, and design can be seen as a means of 'making science visible', in that it is only due to our understanding of the way in which the world works that makes possible contemporary industrial designs. Design science was the attempt to systematise design as a rational decision-making process, with laws of design and rules that could bring design to resemble the positivistic vision of science. Cross appears to distance most design theorists from such a vision of design, with a quote from a major theorist to the effect that design 'is itself a nonscientific or ascientific activity' (p. 53). If design is not scientific, it may still be possible to scientifically study design, and it is this interpretation that drives what Cross terms as the design methodology movement, seeking to understand the 'principles, practices, and procedures' (p. 53) of design. Cross contended that are aspects of the design process still not very well understood, such as the role played by sketching in the ideational process, or what might be better or worse processes in which designers worked. Cross' own assessment circa 2001 was to avoid the issue of assigning design as a science or art completely, instead referring to it as an interdisciplinary discipline, concerned with the 'artificial world and how to contribute to the creation and maintenance of that world' (p. 54).

We return to the question that we began this section with: Is design a science or an art? It would appear that our best response must be: it is both. There are aspects of design which draw from artistic ways of thinking, in that one is required to have a particular 'professional vision' (Goodwin, 1994) that can only be acquired through an extensive private apprenticeship in the discipline. On the other hand, there are also rational, representable 'pieces of knowledge' that one can communicate efficiently through 'public' technologies such as conventional lecture style classroom pedagogy. For schools then, our current understanding of design does not quite yet pose serious challenges: it might be possible, especially at the level of a general public education, to suppose that a general 'appreciation' of design via mass instruction is 'good enough' for a wide range of students. After all, a similar problem exists for science: for public education, a gradually increasing level of supervision qualifies people to work in increasingly privileged circles of scientific work. We expect that by the time someone is prepared to produce new scientific knowledge, they would have some apprenticeship under established scientists, while novices and general consumers of science need only passing familiarity. So far, so rational—such a model can be said to be a consequence of a form of 'operations research', a type of design study, interested in the design of social organisation for particular goals. If one was interested in a particular kind of social outcome, and one had only limited resources, such a line of thinking would probably be, as Herbert Simon would have put it, the choice that satisficed the constraints. Yet, as much as we understand design to be the discipline of artifice, we need to understand that there is no inherentness, no naturalness in the way we organise (public) schooling. It is entirely an edifice, designed to meet certain constraints and produce a particular intended social outcome. The intentions that we might place on the goals of school are political matters, and it is a political sleight-of-hand for any government to position the politics of educational provision under the guise of the supposedly neutral language of 'the economy'. To understand this, we need to consider other aspects of design: the role of intentions in design and the nature of design problems. To prefigure the argument, my position is that the open-ended nature of design problem solving can reveal to many the idea that the current compromises around school need not have its 'natural' appearance that it currently does. Design problems cannot be definitively 'solved' and always involve compromises. Knowing these ideas about design is like providing students with a metaphorical master key to thinking about their place in the social order. Of course, having the master key will not guarantee that one will have the power to open the door, but at the very least, doors will no longer be locked out.

Wicked problems and design

Even from the time when Herbert Simon attempted to scientise design, it was clear that the kinds of problems that design dealt with were different from that of the sciences. While the natural sciences dealt with problems of analysis in close-ended, bounded problems, designers dealt instead with open-ended problems whose solutions could often involve what philosopher and cognitive scientist Daniel Dennett (2013) referred to as 'Jumping out of the system' ('JOOTSing'). This term refers to the way in which systems of thought inevitably possess rules for its proper functioning. The essence of creativity is to be found in the deliberate violation of the rules in order to create something new; as a consequence, JOOTSing is not a simple matter of randomised novelty, but a considered response to a system and its internal rules, and requires that practitioners be deeply immersed in its rules. It is for this reason that design is often not taught at an early stage of general education as most students would still be learning about the general rules of the systems that they might be designing. To the extent that Herbert Simon's attempt at creating a science of design was influential, an equally influential response somewhat in opposition to thinking of design as a science came from Horst Rittel (1972) and Rittel and Webber (1973). In an authoritative manner not commonly seen, Rittel and Webber declare even in their abstract that '[t]he search for scientific bases for confronting problems of social policy is bound to fail, because of the nature of these problems' (p. 155). Rittel is famous for his distinction between what he termed as tame and wicked problems: the distinction is not between levels of moral turpitude, but in the degree to which each kind of problem is amenable to solution. Here, the distinction between tame and wicked problems is between difficult problems such as finding solutions to sophisticated mathematical or scientific puzzles, or even planning the optimum sequence of moves in chess versus the kind of open-ended problems such as social planning.

Consider, for instance, policy for the provision of public healthcare: in many countries, its citizens have determined that access to high-quality healthcare is a human right and therefore have instituted universal health coverage to pay for these services. On the other hand, and most notably in the United States, healthcare is perceived as a privilege that one obtains as a matter of one's 'merit', often measured in economistic terms. While socialised access to medicine may be a good idea, the criticism that is often levelled at it is its relatively long wait times for specialised treatments, whereas the private medical model is praised for its efficiency. On the other hand, the sincere wish to avoid fellow human suffering due to circumstances beyond one's control could be worth more than any potential economic gains from making healthcare a profitable venture. We notice a few features of policy problems such as these. Firstly, it can be rather easy for the common layperson to have opinions about how policy ought to be designed. Policy goals can vary, as we can see equally valid goals for healthcare being the optimisation of human flourishing, in terms of public health, or private wealth, or some combination in between. The choice of framing thus produces its own set of problems, as policy solutions that optimise for certain goals will often do so at the expense of other goals. Such complex open-ended problems should be contrasted against closed-ended ones, where the problem boundaries are clearly demarcated, possible solutions can be easily recognised to be correct or wrong, and solutions, once found, will end the search for further solutions. Take, for instance, a hard problem such as the search for an optimum strategy for the playing of games like chess or Go: these problems have very well defined rules, win and loss states, and end points. Even though the rules guarantee that the branching logic that unfolds as the game progresses results in a large number of possible moves, all these moves are in theory finite in extent. Most importantly, at every step of the game, the next best move or set of moves is always clear. With wicked problems, it is seldom clear if one or another course of action is ideal. Even the search for explanations for complicated natural phenomena is aided by the fact that solutions are bounded by the requirement to explain phenomena. Much of the character of wicked problems arises due to the open-ended nature of human intention and agency and also due to the fact that what is considered ideal by humans changes across time and space; cultural preferences vary over what can be rather arbitrary specification of preferences, and different cultures privilege different things.

This results, as Rittel and Webber (1973) surmised, in ten features of wicked problems:

1. There is no definitive formulation of a wicked problem.
2. Wicked problems have no stopping rule.
3. Solutions to wicked problems are not true/false, but good/bad.
4. There is no immediate and no ultimate test of a solution to a wicked problem.

5. Every solution to a wicked problem is a 'one-shot operation'; because there is no opportunity to learn by trial and error, and every attempt counts significantly.
6. Wicked problems do not have an enumerable (or exhaustively describable) set of potential solutions, nor is there a well-described set of permissible operations that may be incorporated into the plan.
7. Every wicked problem is essentially unique.
8. Every wicked problem can be considered a symptom of another wicked problem.
9. The existence of a discrepancy representing a wicked problem can be explained in numerous ways; the choice of explanation determines the nature of the problem's solution.
10. The designer has no right to be wrong.

It is clear that the provision of public goods and especially, pertinent to this discussion in this chapter, the design of policies for public schooling are wicked problems. The curriculum for STEM education, for instance, relies on rather arbitrary goals for public schooling. While the conventional purposes of attending to economistic goals appear to be a recurrent justification in societies all over the world, such apparently 'neutral' goals can easily be shown to serve dominant interests, often exacerbating inequalities in the process. Feature 1 of wicked problems suggests that while a curriculum for STEM education may serve 'the economy', it could equally well be directed for the furtherance of liberatory goals in societies.

Feature 2 is apparent: there is unlikely to be a curriculum document today that will be sufficient 20 years hence. Feature 3 is also apparent: STEM curricula can only be judged as better or worse fits for the problems that they intend to solve. Feature 4 arises as a result that we cannot know the results of a STEM curriculum implementation until many years have passed. Standardised tests may reveal short-term outcomes, but it is only over the long term that we can see if students are adequately prepared to live their lives as economic and political agents of their time. Even then, the lack of ultimate test feature arises because of the complexity of the interactions involved in education: if a policy 'works' how might we be able to know that it is truly the result of the policy, and not, say, the changing economic tides set in place by forces beyond the control of the state? Feature 5 suggests that all wicked problems are one shot; McCall and Burge (2016) suggest that this feature is probably mistaken; I will discuss this below. Feature 6 is obvious as well: curriculum proposals for STEM can have a wide range of possible goals, not all of which may be specifiable from the outset. Feature 7 arises as a consequence of the cultural differences that school policy environments have to operate in. Feature 8 is clear when we consider that schooling policy is the result of attempting to solve the wicked problem of social planning for the future. Feature 9 again is self-evident from Feature 1. Lastly, Feature 10 places responsibility squarely on the designer: given the problem conditions, designers have the obligation to do their best to arrive at solutions that are as good as they can get it. Designers have no right, despite the nature of wicked problems, to excuse their design solutions as not good enough on the grounds that it is not possible to have a solution that is good enough.

To be sure, McCall and Burge (2016) remind us that there are some caveats to these features of wicked problems. For instance, Feature 5, the 'one-shot' nature of wicked problems, can be clearly untrue if one considers that prototypes for design problems are possible. Software design, for instance, often considers how actual users will make use of software by actually deploying versions of software and then monitoring usage while at the same time developing further iterations of the software. Regardless of these caveats, the concept of design as a method interested in wicked problems is incredibly productive and continues to be used even in contemporary times, and in fields such as policy analysis (Peters, 2017), and even technological deployment in education (Borko, Whitcomb, & Liston, 2008). Richard Buchanan (1992) proposes that while the list of 10 features proposed by Rittel serve a good purpose in illustrating the unique nature of design problems, how these features come about as a consequence of the manner in which design is practiced is not well understood. Buchanan proposes that the unique nature of design problems arises from the major distinction that while the natural sciences and practically all forms of disciplinary knowledges are interested in abstracting from the particular towards the formation of generalisable representations, designers are actually interested in going the other way: 'The subject matter of design is potentially universal in scope, because design thinking may be applied to any area of human experience. But in the process of application, the designer must discover or *invent a particular subject* out of the problems and issues of specific circumstances' (p. 16, emphasis added). While a problem in the natural sciences (to take an example of a classically tame problem) may be initially undetermined or under-determined, part of the investigation into finding solutions involves 'pinning down' (as it were) these indeterminacies such that a consistent problem may be formulated for generalisable solution. On the other hand, design deals with two levels of knowledge, a general level that deals with the nature of the ideal design, and a specific level which deals with the instantiation of these general principles in a particular artefact.

The notion that design problems are wicked problems buys us significant insight into the educative purposes makerspaces and STEM can serve. In contrast to a linear, mechanistic model of the search for *the* unique solution to a closed and carefully bounded problems, the quest for solutions for wicked problems does not follow a linear trajectory; graphically, we may plot the progress towards finding solutions over time as in Figure 4.1.

Tame problems are the form which we prefer to address through schooling, and especially in STEM classrooms because it appeals to the positivistic notions of how scientific knowledge progresses and how we may bring students through a quantifiable increase of their knowledge. Bringing students through the solution of tame problems feels appropriate for school, as much of schooling has been

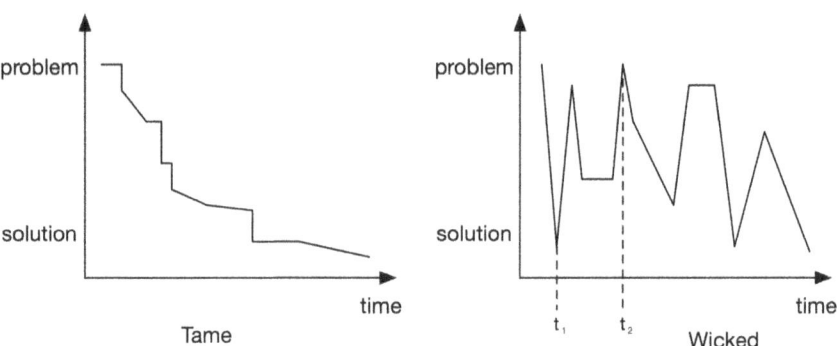

FIGURE 4.1 Metaphorical progression from problem to solution over time—tame versus wicked problems

modelled after a rational, construction-based metaphor of progressive adding, as if knowledge was an object that could be built. In contrast, wicked problems are typically avoided in school situations because no satisfactory end point in the investigation can be obtained. Consider the progress of wicked problem solution finding as in Figure 4.1. When a problem is initially identified, say, the need to construct a bridge to span a ravine, candidate solutions may emerge quickly. The basic idea may be to build a suspension bridge with two piers, and plans may be drawn up, materials sourced and labour secured. At this point, we may be at time t_1, and we may believe that the problem was solved. Upon starting work, we might find that the proposed site is actually archaeologically significant, a fact that was missed in the initial ground survey. This poses new problems: do we desecrate the grounds or shift the bridge site to a location that is less centrally connected to the traffic network? Perhaps an archaeological study is carried out, which seems to solve the problem by extracting the artefacts to a museum, but we may then encounter further problems when the indigenous population protests the proposal to even consider building a road near its hallowed ground. And so on. The point of this is that how 'solved' a problem is depends on what time point one decides to make the assessment. In schooling contexts, where there is more often a product focus in assessment, wicked problems pose a major challenge. Even though a student may have made much effort through the process of working through the problem, they may have only arrived at a problem state (t_2). Assessments that measure the gap between the initial and final state will miss out the processes that students would have been through. To be sure, there is a certain degree to which this problem is exaggerated—student maintained design journals can go a long way towards recording and recognising intermediate insights and work done.

On a more serious note concerning the kinds of problems deemed worthy for school, the gatekeeping function of school demands assessment methods which can justifiably be considered fair. Wicked problems, which accept a wide range of possible solutions, not all of which may be reliably compared on a common standard, pose a problem for school systems which desire such forms of assessment. Because of the open-ended nature of design problems, what is deemed as an ideal solution is often a function of the intention of the designer. Let us next consider the role of intention in design.

Design and intentions

A significant motivation for Rittel and Webber (1973) was to dispute the then hubristic assumptions of systems analysts. The systems analysts positioned themselves as universal problem solvers, who, with:

> arrogant confidence, the early systems analysts pronounced themselves ready to take on anyone's perceived problem, diagnostically to discover its hidden character, and then, having exposed its true nature, skillfully to excise its root causes. Two decades of experience have worn the self-assurances thin. These analysts are coming to realize how valid their model really is, for they themselves have been caught by the very same diagnostic difficulties that troubled their clients.

The planning systems that were popular at the time needed its designers to become clear of the desired outcomes, something that Rittel and Webber noticed that was an 'extraordinarily obstinate task' (p. 157). This was especially so in social planning situations where goals were contentious. If it were acceptable to design social systems such as healthcare or schooling for a community which one was perceived to be a part of, it would have been quite something else to design for or on behalf of another community, especially if there are significant differences in what each party values. Rittel and Webber pointed out that "the most difficult question of all [was]: 'What *should* these systems do?'" (p. 157, emphasis in original). While Rittel and Webber referred to design problems in policy, there are adequate reasons to suppose that the question of 'oughtness' is one that should occupy designers of all stripes. Designers, as Rittel and Webber contend, are experts in 'identifying the actions that might effectively narrow the gap between what-is and what-ought-to-be' (p. 159).

In the design of everyday objects, for instance, a smartphone, we often need reminding that the problems that these devices solve are not problems that are somehow natural and inevitably a part of the human condition. For instance, while humans as a social species desire for interpersonal connection, we have had few problems with making and maintaining connections with the immediate communities around us and never really needed the adulation of millions of strangers, narcissists notwithstanding. It took the deliberate decision of designers to exploit this desire for connection, inventing new social spaces of interaction, new rules of engagement, and new ways of capitalising and funding, to give humanity the problem that many young adults take for granted. The question that seldom gets asked is the question of whether and how such courses of action are desirable, in no small part because the capitalist motive for profit often trumps any deliberation as to the ethical desirability of particular courses of action. It is often considered that the 'invisible hand of the market' ought to decide desirability, in that design proposals thrive or wither depending on how much designers cater to 'consumers' desire. Yet, the perennial problem with such an approach is the degree to which people can be trusted to make decisions that are in their longer term interests. As this is being written, the exit of the United Kingdom from the European Union is being negotiated, and amidst the clamour of the daily news, the Liberal Democratic Party is suggesting that if it manages to secure a government in the next election, it would revoke Article 50, cancelling 'Brexit'. Naturally, commentators have been quick to suggest that such a move is contrary to the democratic spirit of the original referendum that established Brexit, but a counterargument exists that queries the degree to which the original decision has been made on false premises. 'Market forces' are dependent on a public which has infallible and accurate information to base their decision on; seizing the channels of public communication must be perceived as an act of manipulating human behaviour.

Considering the numerous problems that have emerged from the abuse of self-publishing platforms, it may appear that societies are easily manipulated by a certain informational asymmetry—users of designed products may not be aware of latent design intentions that can be carefully hidden from their view. For the case of social media, for instance, where the design intent is to collect vast amounts of very private information to correlate with variables such as the decision-making steps in purchasing items, the view from the individual is very different from that of the system architect. Collecting the data from large amounts of users, system designers can control for statistical variation such that they can reliably *manipulate* groups towards particular purchasing decision, collecting their middlemen's cut, of course.

Given these issues as a backdrop, the ambition of this book becomes clearer: what can, and should, educators do in an approach to science education that is not merely interested in the communication of conceptual knowledge about science to their students? Specifically, what ought to be the responsibility of educators given what we know about the nature of scientific knowledge, and its role in design of technological objects? In makerspaces, what kinds of artefacts ought to be made real? While it is more than likely that many makerspace contexts have no problems simply using simple construction projects as merely a means to secure student engagement, it remains that more can and should be done.

The work of this section is thus to consider the role that intentions play in design, and how an approach to education in makerspaces can and should deal with developing the appropriate kinds of intentions that we may want our designed artefacts and systems to achieve. A role for makerspaces and science education is in the development of skills of 'reading' the design intent of artefacts. For a very obvious example, public furniture is designed in such a way as to make lying down on them impossible or very uncomfortable. Yet, for many people, these design intentions are transparent, in part because they are not the intended (mis)users of these hostile architectures, and also in part because the skill of 'reading' designs is not widespread. To take a more contemporary example, user interfaces for social media and 'free' games on smartphones are designed to be addictive (Fagerström, 2018). Even something as apparently innocuous as the ubiquitous newsfeed contributes to the addiction: because the newsfeed provides a non-deterministic 'reward' for the act of refreshing, users will not know if refreshing the feed will provide the tantalising piece of information that one can next share with one's social circle and achieve 'internet fame' or merely old news that few others would be interested in. This leads to users continuously checking their applications in order to obtain the reward of novelty. Indeed, such interfaces have even been compared to jackpot machines, with the very common user interface element of the spinning arrow likened to the spin of the jackpot machine display drums.

In more general terms, Nelson and Stolterman (2012) suggest that design is necessarily paradoxical in that any design is both 'magnificent and evil'. Considering addictive user interfaces as an example—while the addictive element is clearly undesirable, especially for segments of the population prone to addictive and compulsive behaviours, such interfaces will attract and retain enough users for the social media platform to be of any value to its users in the first place. Nelson and Stolterman refer to evil not in the sense of 'malevolent forces dedicated to the destruction of everything that is good in the world, or counter to the positive presence of God as in many religious traditions' (p. 184). Rather, they refer to the way in which the act of design is a process of creating a bifurcation, a cleavage, or the imposition of a binary onto the 'one, absolute, and supreme Nature' (p. 185). This is not to take an overly spiritual approach to design, but to recognise that all acts of design are assertions of a particular ordering of things, a statement of what is desirable and ought to be made real. An inevitable part of this process is the casting aside of the 'complementary set' of what is desired, and the boundary crossing and maintenance involved in the declaration of desirability that is design. Nelson and Stolterman assert that any design is, by definition, an act of creating, transgressing, or maintaining this boundary between desirability and non-desirability. While in obvious cases such as the social media example, the disruption caused by such acts of design can cause irritation or hostility, we often underestimate the banal forms of design. Design is also associated with a loss of opportunity, as:

> When a design is brought into the world and made real, its very presence excludes other opportunities. The substantial investment of money, energy, material, and time in a new design directly prohibits other attempts to make alternative designs and realities because of lack of resources. This also holds true for more abstract investments, such as pride and status. This is because identity and self-image become invested in a commitment to the new reality emerging as a consequence of the new design's meaningful presence. (p. 187)

The act of design, in its definition, transgression, and maintenance of boundaries, explains and extends the wicked problem observation that social planning, as a subset of design generally speaking, inevitably creates side effects. As there is no such thing as perfect knowledge, the potential always exists that boundary creation will lead to unintended consequences.

If we accept that inherent in design is the collapse of the wide open possibility of an unbifurcated world, our attitudes towards design should possess a more nuanced appreciation for the possibilities and perils of the design method. In contrast more popular approaches to design instruction which tend to celebrate design as a method for generating creative outcomes in an unproblematic manner (e.g., Kelley & Kelley, 2013; *Stanford d.school*, n.d.), it would be more accurate to also convey a critical perspective to design. Instead of celebrating novelty for its own sake, or ingenious means of achieving design goals, questions that should become routine in design and makerspaces should include (but not be limited to): (i) What is the intention of this design? (ii) What are the ideals that have been promoted in this design? (iii) What possibilities have been removed from consideration in this design?

Instruction in design needs to be cognisant of its ethical dimension. Unlike instruction in the natural sciences and mathematics, design more explicitly involves normative decision-making. It should be quite clear that instruction in the engineering principles that enable one to build a remotely piloted aircraft for recreation also enable the creation of bomb laden drones for attacking one's enemies. Both can be highly creative acts, and what distinguishes these two projects is merely the intention of the designer. In most schooling situations, however, the ethical question is much less stark and hence more difficult to address. Nonetheless, these questions do not go away and still require teachers to respond to them, if only in implicit terms. For instance, in response to the 'user needs analysis' aspect of instruction in design problem solving, schools in Singapore have used what have been deemed as obvious contexts of people with disabilities, or the elderly. The design brief to students often challenges them to come up with solutions for these groups. Unfortunately for most students, they do not have sufficient exposure to such individuals; this often leads to caricatured solutions such as virtual tours of green spaces as substitute to outdoor trips to nearby green spaces for the elderly (an actual project submission I judged for an innovation contest for grade 8 students). While the project may seem innocuous enough, it was clear that there was no sarcasm in the team's intention; they were sincere in their belief that technology in the form of virtual reality goggles was an adequate substitute for the holistic experience of being outdoors.

Such a case demonstrates the need for teachers to go beyond simplistic notions of what it means to be innovative—there are numerous technological tools which may be deployed in novel ways; but only so many of them are in service of ends which are desirable. The role of the teacher is not only necessarily to serve as arbiter of what is desired, and what is not, but to also help students through the decision-making process of identifying appropriate goals. Instruction for innovation should include, as with most forms of instruction, components of reading and writing. For design, the problem is made complex by the numerous 'design languages' that are used, from the diverse fields in which design is practiced. How intentions are expressed in clothing design will be different than in, say, internal combustion engine design. Nonetheless, selecting for the design field that is of interest, a basic activity that needs to be carried out should be some variation of a historical retelling of the key intentions in the field, and how these came to be.

Which came first?

In conventional schooling contexts, it can be common to assume that design, technology, and engineering are to be taught after students have sufficient instruction in the 'basic' disciplines of the natural sciences and mathematics. It can be easy to suppose that since some aspects of design principles are explained by the basic disciplines, it would be ideal to have some familiarity with these principles first. This perspective explains why specialised schools of design and engineering tend to be post-secondary institutions, and in general education contexts design is not formally introduced until the secondary level, even though movements are afoot to make changes. For instance, the Next Generation Science Standards in the United States have begun to include engineering as early as the elementary grades, and in many places around the world, the recently coined moniker of 'STEM' has gained great currency among educators and researchers. However, such an approach to thinking about the relationship between science and technology, specifically, that design and engineering are applications of science and mathematics, is mistaken. It may appear to be of little consequence, and educators will be quick to add that there is often a distinction between practitioners' and school versions of the knowledge being taught. However, I want to argue here that this miscommunication has social consequences and misses opportunities for more progressive forms of education that, again, have social costs associated with it. Briefly, this mistake deepens the hands-minds dichotomy which privileges the role of cognition; in societies, this dichotomy leads to social injustice aided by mistaken ontologies, and a widespread misunderstanding of the interplay between the social and the technical.

Here, I want to make a note that I will use the terms design, technology, and engineering interchangeably. Technology is a wide ranging term that encompasses devices, methods, or processes (Brian Arthur, 2009). The interest of engineering is in the fabrication of such technologies. However, all of these activities are united under the banner of design—design as an art or a science of the artificial helps us understand the overall intention of the activity of creating, using, and maintaining the artefacts that we rely upon for our modern existence. While there will be differences between the exact methods used by designers, engineers, and technologists of various stripes, the irreducible difference lies in their interest in artificing. In this regard, it is useful to consider that the first acts of making took place without any real form of understanding of the natural processes that may explain why materials behaved the way they did:

> Objects are a form of knowledge about how to satisfy certain requirements, about how to perform certain tasks. And they are a form of knowledge that is available to everyone; one does not have to understand mechanics, nor metallurgy, nor the molecular structure of timber, to know that an axe offers (or 'explains') a very effective way of splitting wood. Of course, explicit knowledge about objects and about how they function has become available, and has sometimes led to significant improvements in the design of the objects. But in general, 'invention comes before theory'; the world of 'doing and making' is usually ahead of the world of understanding – technology leads to science, not vice versa as is often believed.
>
> *(Cross, 2006, p. 9)*

Whether science leads technology or the reverse has parallels in the way design problems and solutions are considered. While Cross suggests that technology leads science, he tantalisingly leaves behind the qualifier 'in general'. Given contemporary developments in science, and the manner in which popular reporting of such developments tends to accompany wild promises of its technological applications, it may seem rather obvious that science leads technology. However, advocating such an ontology in schools misrepresents the historical precedent, deepens the mind-body dichotomy, and, crucially, sets up the conditions for technological determinism. More significantly, Cross may not have considered more contemporary complicated, multi-part technologies whose developments have been aided by theoretical insights from science. For instance, the possibility of lasers was hypothesised and predicted before working examples were produced. In modern laser printers, vital sub-parts such as the aperture control of the laser (the part responsible for pulsing the laser on and off onto the printing drum) required a switching frequency measured in kilohertz to be practical, something not realisable without insights from our scientific understanding of polarisation of light (Brian Arthur, 2009).

Regardless, for schools, an important consequence of this is that it should be possible to do design without necessarily understanding the science underpinning it. Herbert Simon (1968/1996) posits that design lives at the boundary of artefacts, in organising the internal structure of an object such that its external function behaves according to what is desired. The suggestion here is that there are levels of abstraction that one could selectively use for the construction of our desired artefacts—one might not know how an axe 'works' metallurgically speaking, but that should not stop one from creating a shelter with it. Similarly, one can create a laser printer without necessarily understanding how polarising elements work. In schools, educators should avoid the relatively common practice of teaching the abstract scientific principle first, before asking students to 'apply' these principles in technologies. Instead, beginning at the level of the artefact can be a useful means to introduce the phenomena, to increase the relevance of the abstract concept, and to increase student engagement. Metaphorically, if we consider Simon's notion of artefacts consisting of an 'inside' and an 'outside', the scientific move is one of digging 'in' through the skin of the artefact in order to understand how it works, while design is a means of combining artefacts and concepts to create newer, 'outer' layers of abstraction. If we are interested in design, all we need to know are the external interaction parameters of the artefacts that we work with—axes or optical shutters—and a plan to combine these, abstracting away how the inner levels of these devices work. To be sure, if one seeks *improvements* in the design of artefacts, however, some degree of understanding the inner layers will become important.

Again, in opposition of design as 'application' of science, we are reminded of the mind-body dichotomy from the previous chapter, and current approaches to learning have demonstrated that there the purely intellectual approach to learning is mistaken because of the nature of human knowledge as metaphorical extension of embodied experiences. The concern in this chapter, however, extends past psychological considerations of critique of the intellectual approach to also consider how such a misrepresentation affects societies. In another sense of 'which came first' which is the theme of this section, I want to ask here about the tension between technologies and societies, with the specific question as to whether technologies drive societal change, or societal needs drive technological change.

Technological determinism is the concern here, and according to Wyatt (2008), it has two major interpretations that arise from creating and maintaining the false dichotomy between technology and societies. In the first sense of technological determinism, the belief is that technologies stand apart from societies: here we get the myth of the heroic lone inventor working in laboratories both physically and cognitively distant from societies. In the second sense, which Wyatt asserts to be more significant, the belief is that technologies can change societies and in fact leads to social change. Stronger forms of technological determinism may insist that technologies are the driving factor for social change, and hence we arrive at naming particular historical periods after the technologies that have been prominent: stone, bronze, iron, steam, and more recently, information.

Technological determinism is behind bold claims that railways in India would eliminate the caste system or that electrification would lead to communist utopia, or, for an example which is relevant to this book, new learning technologies will change the way people learn. Such overblown claims have led well-placed observers to issue strongly worded criticisms about the relationship between technology and education (Bayne, 2015; Selwyn, 2016). We are apt to look at recent events such as the 'Arab Spring' protests and revolutions and be swayed by pronouncements that this has been the 'Facebook revolution', because much of the organising and coordination has been done over social media and other digital media. What is less often told is the already existing human organisations that were amplified with technology in places such as Egypt and Syria, while technology was essentially impotent in other states such as the United Arab Emirates where civil society organisations had been weak (Toyama, 2015). Indeed, such a faith in the ability of technology to change the world has been the subject of numerous recent critiques (e.g., Carr, 2014; Edgerton, 2006; Morozov, 2013), which in one form or another are demonstrations of the harms that emerge from an excess of technological determinism.

Looking back at this point, Chapter 2 of this book had been a critique on the binarism of power and knowledge—with positivism taking the position of the superiority of certain forms of knowledge, and postpositivism taking the opposite stance that power is inseparable from knowledge. Chapter 3 discusses the mind-body dualism. In both, the difficult position that we are left with is that there any insistence on either pole of the binary distinction is mistaken. Knowledge is socially constructed, however, there are ways to discern how some knowledge claims are better than others. We learn using our minds *and* our bodies. In this chapter then, the binarism that I want to address is that of the distinction between societies and technology. As Wyatt (2008) surmises for researchers in science and technology studies (STS):

> The challenges for STS remain: to understand how machines make history in concert with current generations of people; to conceptualize the dialectical relationship between the social shaping of technology and the technical shaping of society [...] These dialectics are unresolvable one way or another, but that is as it should be. What is important is to continue to wrestle with them. We need to take seriously the efforts to stabilize and extend the messy and heterogeneous collections of individuals, groups, artifacts, rules, and knowledges that make up our sociotechnical world. We need to continue to grapple with understanding why sometimes such efforts succeed and sometimes they do not. Only then will people have the tools to participate in creating a more democratic sociotechnical order. (p. 176)

Makerspaces, as sites for learning about making, designing, the sciences and technologies that drive contemporary life, should provide opportunities for students to learn about the holistic interactions of science, technologies, its influences on, and its reliance upon societies. This is potentially a tall order for science educators merely looking at makerspaces as sites to interest students with 'fun' making activities that can introduce students to the scientific content that is needed for their passing the next gatekeeping goal. Most educators would complain that the school should not be a site of political contestation; science educators especially are prone to thinking of science and technology as politically neutral and distant from the conflicts of society. Yet such a position plays directly into the technological deterministic view, misrepresenting the complex interactions. I will say more of this in the later section, but there is a final point here regarding the interplay between societies and technologies.

Quite often, technological determinism makes obscure the relationship between technology and society—if we wish to explain the rise of social media, for instance, technological determinists would reach towards ideas such as the arrival of particular technologies such as the open internet and encrypted communications. Yet, these explanations will likely miss out the sociopolitical dimensions such as the massive initial investment (in the United States) into a redundant form of communication in case the cold war with the Soviet Union ever became a hot one. Similarly, observers of the Californian economy are likely to point to the oversized effect of monetising new technologies and miss the systemic contradictions such as rampant and entrenched class, race, and gender discrimination despite widespread proclamations of support for the contrary (Barbrook & Cameron, 1996). The point here, following Wyatt (2008), is that societies and technologies are irretrievably codependent on each other, and explanations of the success *and* failure of particular courses of action or technological implementation should abide by the principle of symmetry first advocated by Bloor (1991, in Wyatt, 2008). Here, the same *sociotechnical* explanations ought to be used for both forms of explanations, it is completely ingenuous to, for instance, attribute the success of some technological implementations to its technical superiority, whereas another failure is attributed to social causes.

For an example pertinent to our interests here, just as much as one would like to suppose and demonstrate empirically that makerspaces, as sites of 'cool new tech', are effective means of science and technology instruction, we need to both (i) consider the sociocultural factors that contributed to the technological adoption and success, and alternatively (ii) not consider failed examples of makerspaces as merely sites where the people 'failed'. In other words: makerspaces should have a focus on both people *and* things. It is not enough to merely focus on the technical aspects of the things that make interventions work, as that would be a rather severe form of reductionism, and potentially ignores the Hawthorne effect. Neither is it enough to only make use of sociocultural explanations for the success or failure of makerspaces.

To summarise this section, there are two main points I wish to make here: firstly, as a slightly more minor point, while schools often represent design, technology, and engineering as 'applied' versions of knowledge in the natural sciences and mathematics, it would rather be more accurate to understand that technologies are logically prior to the sciences, simply because at least for simple, discrete technologies, we have always been able to accomplish things without an understanding of the natural principles that explain it. On the other hand, if we begin to consider more complicated technologies, its growth can be (and often is) aided by insights from the natural sciences. Nonetheless, there can be layers of abstraction possible such that understanding the underlying layers may not be necessary for the operation of the upper layers. Considering the relationship between societies and technologies, it is important that we do not ascribe too much of the account to either. Instead, these terms exist in a dialectical relationship, each influencing the other inseparably. The educational implications of this section derive from a deeper consideration of the relationship between science and technology/design/engineering: instead of considering design as subsidiary and derivative to the sciences, we need better ways of presenting the interrelationship. Further, the relationship between technologies and societies needs better representation in the classroom. If Chapter 2 can be interpreted as a social epistemology of the natural sciences, a significant focus here is on the social epistemology of technology. I want to argue here that, just as science and societies are related by the manner in which truth claims are made and adjudicated, the truth and falsity of technologies are also a matter of societal consideration. Here, technologies are 'true' in the manner in which they succeed or not in being widely adopted. This line of thinking buys us in education an alternative means to judge the goodness of designs—whether or not proposed designs are more or less likely to attend to social needs, and whether appropriate social needs have been identified. More than merely considering the aesthetic value of design for the purposes of adornment and increasing its economic appeal, design requires us to consider sociological and anthropological factors, in addition to the scientific/ technological. These considerations lead us into thinking about how school systematically misrepresents the relationship between science, technology, people, and societies, and this will be the theme of the next section.

What is wrong with school?

At least in countries that have been influenced by the British and American models of public schooling, the dominant role of school for societies has been the efficient preparation of workers for the needs of industry. From this perspective, capitalists' investment in communities and societies in the form of compulsory taxation is returned in a 'workforce' that is suitably skilled in the techniques required for economic activity. To be sure, other goals are often present; the creation of cohesive societies and the attempt to solve all manner of social issues have often been saddled upon schools (Labaree, 2008), often in ways that overestimate the power of schools to change society (Apple, 2002; Labaree, 2016). Specific to science and technology instruction, the predominantly technical nature of these disciplines usually obscures its sociological relevance; at least in the United States, there have been a seemingly endless stream of calls for better science and technology instruction, not for social ends, but for ideological and economic goals. For instance, science and technology instructions in the 1960s were gripped with a desire to defend against the former Soviet Union in the wake of the placing into orbit of the Sputnik satellite; in the 1980s, the Asian 'Tiger' economies posed such a threat that politicians' rhetoric went into overdrive, accusing opponents of 'unilateral education disarmament' (Bracey, 2003; NCEE, 1983). Such rhetoric has been refreshed in recent years (Gallagher, 2013). Even for making and makerspaces, President Obama's opening of the White House Maker Faire (White House, 2014) was, unsurprisingly, accompanied by economic rationales for making.

The rhetoric from the National Commission on Excellence in Education (1983) is so striking and borderline hyperbolic that it is worth repeating here in full, just so that we can become alert to its echoes over 30 years later:

> Our Nation is at risk. Our once unchallenged preeminence in commerce, industry, science, and technological innovation is being overtaken by competitors throughout the world. This report is concerned with only one of the many causes and dimensions of the problem, but it is the one that undergirds American prosperity, security, and civility. We report to the American people that while we can take justifiable pride in what our schools and colleges have historically accomplished and contributed to the United States and the well-being of its people, the educational foundations of our society are presently being eroded by a rising tide of mediocrity that threatens our very future as a nation and a people. What was unimaginable a generation ago has begun to occur—others are matching and surpassing our educational attainments [...] If an unfriendly foreign power had attempted to impose on America the mediocre educational performance that exists today, we might well have viewed it as an act of war. As it stands, we have allowed this to happen to ourselves. We have even squandered the gains in student achievement made in the wake of the Sputnik challenge. Moreover, we have dismantled essential support systems which helped make those gains possible. We have, in effect, been committing an act of unthinking, unilateral educational disarmament.

Upon this background, and widespread agreement among governments internationally of the economic (and national defence) value of schooling, programmes of international quantitative comparison must be perceived as inevitable outgrowths. We should not forget that the Program for International Student Assessment (PISA) is managed by the Organisation for *Economic* Cooperation and Development, and not, for instance, the United Nations Educational, Scientific, and Cultural Organisation. While it can plausibly be argued that tests such as the PISA can be made to measure more than economically relevant indicators, criticisms of quantitative international comparisons have focussed on the reductionist nature of these assessments. For instance, Morris (2015) points out that factors other than education influence economic success, even as advocates of international comparison deploy crisis rhetoric and suggest that the correlation between educational and economic success is somehow causational. Labaree (2014) critiques quantitative comparisons as either measuring 'mastery of skills that are relevant but not taught [PISA] and the other on mastery of content that is taught but not relevant [NCLB[1]]'. Luke (2011) reminds us that while standardisation works for systems such as shipping containers and harmonising international trade, educational success is a complex phenomenon that depends on, and leads to, many interrelated factors and should not be reduced to simplistic measures.

In all of these critiques, the crux of the matter is the manner in which we utilise reductionism (Allen, 1991; Reich, Garrison, & Neubert, 2016) to handle the complexity of phenomena. The educational interaction in school, on the face of it, primarily happens at the interface of the teacher and the student. However, any conscientious parent would certainly object to the notion that their involvement with their child's education stops at merely transporting their child to and from school. Indeed, a complete treatment of the phenomena of schooling can consider all the 'layers' of analyses, from the neurological, through the psychological, all the way to the sociological, and certainly the political especially if we consider questions of curriculum. Schooling is a complex phenomenon and in many ways is irreducibly so. Unfortunately, because of the scientism dominating our societies, we often desire to make 'scientific' claims about systems which cannot be reduced. Our understanding of Economics, for instance, does not extend past fairly generic and simple situations; our models are limited in its explanatory and predictive value (Luyendijk, 2015). Similarly, for education, this desire to produce 'scientific' knowledge about its processes and products, especially in response to pressures from governmental pressures to see greater 'return on investment' in public schooling, has seen numerous rational sounding proposals based ultimately on reductions in complexity to make the problem tractable. Hence, for instance, many leading journals of (science) education research have predominantly adopted a programme evaluation perspective, repeatedly showing the efficacy of various educations interventions, but often only implicitly assuming that 'confounding variables' either have been controlled for or can be determined to play only a marginal role.

Even if such studies are well conducted and have justified conclusions, it seems that few users of such forms of research are aware of the notion of wicked problems and how proposals for change often carry with them problems that will emerge as a direct consequence. It is here that we need the wisdom of design as a means for holistic understanding and deliberate change. Instead of a reductionist, scientistic approach, seeking 'active ingredients', 'point masses', or 'genes for X', seeing the problem of school from a design perspective buys us the much needed wider and deeper points of view. With specific relevance to the problem of makerspaces as sites for science and technology instruction, I am therefore reluctant to engage in specific recommendations of particular forms of activity that may have been shown to work in particular context. Instead, following the methods of design, and recognising that education is a design problem, the goal of a book such as this is to inform and initiate discussion on the level of intentions and goals. Take, for instance, the design of chairs, as a class of surfaces for seating on. While there is no general theory of chairs, no discipline of 'chair-ology', there remain some ideals that for chair-ness that one may specify. Such ideals, desiderata, may be distinct for chairs than for couches, but these are certainly not to be confused with the specifics such as seat height, joint angle, surface texture, or any other parameter that describes the specific implementation of a chair. Unlike a possible science of the relationship between seat back recline angle and productivity on computer usage (for example), the ideals of design are rather more ephemeral, concerned with ideals such as if, in the first place, a chair is an ideal resting surface, and what qualities of chair-ness should designers of specific instances of chairs be concerned with, and why.

From this perspective, we should desire makerspaces to attend to certain problems of the presentation of science and technology in the school, in no small part because of the nature of the problems that we face in our contemporary age. The reports of the United Nations Intergovernmental Panel on Climate Change serve all of humanity notice that we cannot continue 'business as usual'. The standard conveyer belt that brings students through school as industrial production line for bolting on knowledge in preparation for economic participation needs a significant reconsideration, especially if current economic models are contingent on cheap resources, unjust social relationships, and boundless sources of energy (Patel & Moore, 2017). We need new ways of being, new goals for societies, and new ways of productively occupying individuals in tasks that do not lead, either directly or indirectly, to the destruction of the very things that we need for our long-term survival as a species. Such needs require tremendous amounts of innovation, and while this chapter has not mentioned much about this aspect of design (I will do so in the next), it should be sufficient here to say that our old ways of thinking are not likely to give us the solutions we need. This is a very delicate situation indeed: we do not wish to throw away the baby with the bathwater. There have been significant gains in mass public literacy with our current approaches to public schooling. Yet, the social environment has changed while schools remain largely conservative institutions, with the well-worn aphorism that an 18th century school teacher would feel at home in contemporary classrooms repeatedly being used by 'ed-tech' reformers every few years.

The aphorism is wrong and right, of course: much of the content and psychological tenor of schooling have changed; however, the organisational structures, and aims and purposes of schooling, have not moved very much. As noted above, the appearance of artefacts such as 'technology' in the classroom should not be confused with any fundamental form of change. We should regularly ask ourselves if such changes are actually improvements (Biesta, 2016); the concern here is that we are designing better mousetraps while what we actually need is a means to address the termite infestation that threatens to bring the house down. A significant carryover that I want to address here is the concept of disciplinary distinctions and its possible role in perpetuating the crisis that we are encountering. Not surprisingly, the idea of academic disciplines is rooted in reductionism, in the tendencies of societies influenced by the very Western notion that there are primary and secondary qualities and that the primary qualities are to be privileged over the secondary (Allen, 1991). Primary qualities refer to shape, size, mass, number, and so on; while secondary qualities refer to colour, taste, sound, smell, and so on. Under reductionism, the desire for 'objective' measures, and the suspicion of subjectivity demanded by the earlier interpretations of modern science, it becomes inevitable that academics and educators have sought to distinguish forms of knowledge into its 'pure' forms. Schooling becomes an exercise in acquiring 'facts' and not about values:

> One general effect of Reductionism is the elimination of value from the world which is then interpreted 'objectively' as a realm of bare and neutral facts. On this basis, all subjects should be taught in a merely factual way, as bodies of information and sets of manipulative skills, and not also as matters of values which are to be appreciated, admired, and cherished. Education, thus, becomes a matter of imparting only knowledge about things and of developing only techniques of manipulating them, of the head and the hand and never the heart.
>
> *(Allen, 1991, p. 26)*

Nowhere is this situation more acute than in scientific instruction. There have been research attempts at teaching science within the context of societal concerns, such as the movements of Socio Scientific Issues (Sadler, 2011; Zeidler, Sadler, Simmons, & Howes, 2005); Science, Technology, Societies, and the Environment (Pedretti & Nazir, 2011); and Socially Acute Questions (Morin, Simonneaux, & Tytler, 2017). However, these approaches have not been widely taken up by school systems, in no small part because science teachers in many places around the world have been trained in reductionist, discipline centred ways of looking at the world. Considering the effect of science and technology on societies and the environment is certainly complicated by the fact that in most popular understanding of science and its effects on societies, scientific knowledge is often considered the primary quality while the contextual nature and messiness of social values are rendered as irrelevant to science instruction.

A way forward

A proper appreciation of the knowledge of design, and especially the critique of technological determinism, ought to be a central part of the school curriculum. Science and technology (and other forms of knowledge) have been remarkable cultural achievements and mostly deserve reproduction as forms of knowledge that help us understand the world. However, an overemphasis on cultural reproduction can stifle our collective possibilities for thinking otherwise, essentially robbing our children of their right to an open future (Feinberg, 1994) even though openness can be hard to discern (Mills, 2003). Our cultural achievements exist within a historical context, and there are associated contingencies that cannot, and should not, be deleted. It has been a brilliant intellectual sleight of hand for the sciences to have asserted science as a knowledge that exists outside of space and time. Yes, we have had spectacular successes, including the sending of space probes beyond the edge of our solar system, but we have also many embarrassments such as sincerely held beliefs about eugenics and anthropometry (Gould, 1981). Perhaps to a certain degree of hyperbole, it is now, more than ever, that we need new and better ways forward. A different form of science education that understands and communicates the social nature of truth claims, the social nature of technology, and the embodied nature of learning needs to emerge. Design, as interested in the particular, as an interdisciplinary study of how and why intentions should be made real, stands before us as a candidate, mostly misunderstood within our schooling contexts obsessed with ahistorical decontextualised forms of knowledge. The role of makerspaces within such a framework of goals for schooling is as a site for contextualisation. Makerspaces fight the desire to reduce phenomena to mere representation. Representations attempt but will never express the fullness of the phenomena; makerspaces are sites of demonstration, decomposition of technologies to its constituent components so as to discern how they work, and crucially, the construction of alternative technologies that are driven by intentions other than what we currently possess.

Note

1 The No Child Left Behind policy prescribed regular standardised testing to assess system performance.

References

Allen, R. T. (1991). Reductionism in education. *Philosophical Inquiry in Education*, *5*(1), 20–35.

Apple, M. W. (2002). Does education have independent power? Bernstein and the question of relative autonomy. *British Journal of Sociology of Education*, 607–616. https://doi.org/10.1080/0142569022000038459.

Barbrook, R., & Cameron, A. (1996). The Californian ideology. *Science as Culture*, *6*(1), 44–72. https://doi.org/10.1080/09505439609526455.

Bayne, S. (2015). What's the matter with "technology-enhanced learning"? *Learning, Media and Technology*, *40*(1), 5–20. https://doi.org/10.1080/17439884.2014.915851.

Biesta, G. (2016). *The beautiful risk of education*. Abingdon: Routledge.

Bloor, D. (1991). *Knowledge and social imagery* (2nd ed.). Chicago, IL: University of Chicago Press.

Borko, H., Whitcomb, J., & Liston, D. (2008). Wicked problems and other thoughts on issues of technology and teacher learning. *Journal of Teacher Education*, *60*(1), 3–7. https://doi.org/10.1177/0022487108328488.

Bracey, G. W. (2003). April foolishness: The 20th anniversary of a nation at risk. *Phi Delta Kappan*, *84*(8), 616–621.

Brian Arthur, W. (2009). *The nature of technology*. London, UK: Allen Lane.

Buchanan, R. (1992). Wicked problems in design thinking. *Design Issues*, *8*(2), 5–21.

Capraro, R. M., Capraro, M. M., & Morgan, J. R. (Eds.) (2013). *STEM project-based learning: An integrated science, technology, engineering, and mathematics (STEM) approach*. Rotterdam: Sense Publishers.

Carr, N. (2014). *The glass cage: How our computers are changing us*. New York, NY: W. W. Norton & Company.

Cross, N. (2001). Designerly ways of knowing: Design discipline versus design science. *Design Issues*, *17*(3), 49–55. https://doi.org/10.2307/1511801.

Cross, N. (2006). *Designerly ways of knowing*. London: Springer-Verlag.

Dennett, D. (2013). *Intuition pumps and other tools for thinking*. New York, NY: Allen Lane.

Edgerton, D. (2006). *The shock of the old: Technology and global history since 1900*. London: Profile Books Ltd.

Fagerström, A. (2018, August 10). App Addiction—the Invisible Plague. Medium; UX Planet. https://uxplanet.org/app-addiction-the-invisible-plague-73447926d734.

Feinberg, J. (1994). *Freedom and fulfillment: Philosophical essays*. Princeton, NJ: Princeton University Press.

Feyerabend, P. (1988). *Against method* (Rev., p. viii, 296). New York, NY: Verso. (Original work published 1975.)

Gallagher, J. J. (2013). Educational disarmament, and how to stop it. *Roeper Review, 35*(3), 197–204. https://doi.org/10.1080/02783193.2013.799412.

García Martinez, A. (2016). *Chaos monkeys: Inside the Silicon Valley money machine*. London: Ebury Press.

Goodwin, C. (1994). Professional vision. *American Anthropologist, 96*(3), 606–633.

Gould, S. J. (1981). *The mismeasure of man* (1st ed., p. 352). New York, NY: Norton.

Guilford, J. P. (1950). Creativity. *The American Psychologist, 5*(9), 444–454. https://www.ncbi.nlm.nih.gov/pubmed/14771441.

Jones, J. C. (1997). How my thoughts about design methods have changed during the years. *Design Methods and Theories*, 11, 1.

Kelley, D., & Kelley, T. (2013). *Creative confidence*. Random House.

Kuhn, T. S. (1996). *The structure of scientific revolutions* (3rd ed.). Chicago, IL: University of Chicago Press. (Original work published 1962.)

Labaree, D. F. (2008). The winning ways of a losing strategy: Educationalizing social problems in the United States. *Educational Theory, 58*(4), 447–460. https://doi.org/10.1111/j.1741-5446.2008.00299.x.

Labaree, D. F. (2014). Let's measure what no one teaches: PISA, NCLB, and the shrinking aims of education. *Teachers College Record, 116*(9), 1–14.

Labaree, D. F. (2016). An affair to remember: America's brief fling with the university as a public good. *Journal of Philosophy of Education, 50*(1), 20–36. https://doi.org/10.1111/1467-9752.12170.

Latour, B. (1986). *Laboratory life: The construction of scientific facts*. Princeton, NJ: Princeton University Press. (Original work published 1979.)

Luke, A. (2011). Generalizing across borders: Policy and the limits of educational science. *Educational Researcher, 40*(8), 367–377. https://doi.org/10.3102/0013189X11424314.

Luyendijk, J. (2015, October 11). Don't let the Nobel prize fool you. Economics is not a science. *The Guardian*. http://www.theguardian.com/commentisfree/2015/oct/11/nobel-prize-economics-not-science-hubris-disaster.

McCall, R., & Burge, J. (2016). Untangling wicked problems. *Artificial Intelligence for Engineering Design, Analysis and Manufacturing: AI EDAM, 30*(2), 200–210. https://doi.org/10.1017/S089006041600007X.

Midgley, M. (1990). The use and uselessness of learning. *European Journal of Education, 25*(3), 283–294. https://doi.org/10.2307/1503318.

Mills, C. (2003). The child's right to an open future? *Journal of Social Philosophy, 34*(4), 499–509.

Morin, O., Simonneaux, L., & Tytler, R. (2017). Engaging with socially acute questions: Development and validation of an interactional reasoning framework. *Journal of Research in Science Teaching, 54*(7), 825–851. https://doi.org/10.1002/tea.21386.

Morozov, E. (2013). *To save everything, click here: The folly of technological solutionism*. New York, NY: Public Affairs.

Morris, P. (2015). Comparative education, PISA, politics and educational reform: A cautionary note. *Compare: A Journal of Comparative and International Education*, *45*(3), 470–474. https://doi.org/10.1080/03057925.2015.1027510.

NCEE. (1983). In National Commission on Excellence in Education (Ed.), *A nation at risk: The imperative for educational advance.* Washington, DC: U.S. Department of Education.

Nelson, H. G., & Stolterman, E. (2012). *The design way: Intentional change in an unpredictable world: Foundations and fundamentals of design competence* (2nd ed.). Cambridge, MA: MIT Press.

Patel, R., & Moore, J. W. (2017). *A history of the world in seven cheap things: A guide to capitalism, nature, and the future of the planet.* Berkeley, CA: University of California Press. https://www.ucpress.edu/ebook.php?isbn=9780520966376.

Pedretti, E., & Nazir, J. (2011). Currents in STSE education: Mapping a complex field, 40 years on. *Science Education*, *95*(4), 601–626. https://doi.org/10.1002/sce.20435.

Peters, B. G. (2017). What is so wicked about wicked problems? A conceptual analysis and a research program. *Policy and Society*, *36*(3), 385–396. https://doi.org/10.1080/1 4494035.2017.1361633.

Reich, K., Garrison, J., & Neubert, S. (2016). Complexity and reductionism in educational philosophy—John Dewey's critical approach in "Democracy and Education" reconsidered. *Educational Philosophy and Theory*, *48*(10), 997–1012. https://doi.org/10. 1080/00131857.2016.1150802.

Rittel, H. (1972). On the planning crisis: Systems analysis of the "first and second generations." *Bedriftsøkonomen*, *8*(Norway), 390–396.

Rittel, H. W. J., & Webber, M. M. (1973). Dilemmas in a general theory of planning. *Policy Sciences*, *4*(2), 155–169. https://doi.org/10.1007/BF01405730

Sadler, T. D. (Ed.). (2011). *Socio-scientific issues in the classroom: Teaching, learning and research.* London: Springer.

Selwyn, N. (2016). Minding our language: Why education and technology is full of bullshit … and what might be done about it. *Learning, Media and Technology*, *41*(3), 437–443. https://doi.org/10.1080/17439884.2015.1012523.

Simon, H. A. (1996). *The sciences of the artificial* (3rd ed.). Cambridge, MA: MIT Press. (Original work published 1968.)

Stanford d.school. (n.d.). Stanford D.school. Retrieved June 19, 2019, from https:// dschool.stanford.edu/.

Tan, M., Lee, S.-S., & Ng, Z. Y. (2017). Social influences on student perceptions of failure in learning design processes: Instructional implications. *Learning: Research and Practice*, *3*(2), 130–147. https://doi.org/10.1080/23735082.2017.1351577.

Toyama, K. (2015). *Geek Heresy: rescuing social change from the cult of technology.* New York, NY: Public Affairs.

White, H. (2014). *The first white house maker faire.* The White House. https://obamawhite-house.archives.gov/node/316486.

Wyatt, S. (2008). Technological determinism is dead: Long live technological determinism. In E. J. Hackett, O. Amsterdamska, M. Lynch, & J. Wajcman (Eds.), *The handbook of science and technology studies* (pp. 165–180). Cambridge, MA: MIT Press.

Zeidler, D. L., Sadler, T. D., Simmons, M. L., & Howes, E. V. (2005). Beyond STS: A research-based framework for socioscientific issues education. *Science Education*, *89*(3), 357–377. https://doi.org/10.1002/sce.20048.

5

THE USES AND ABUSES OF SCIENCE AND TECHNOLOGY

Continuing the theme that was developed in the last chapter, I want to further develop in this chapter the social aspects of science and technology. Specifically, this chapter will deal with the question of whether (and how) science and technology is good for society, and more generally, the inter-relationships between science, technology, and societies. This chapter is important in two significant ways. Firstly, in developing a curriculum for science, technology, engineering, and mathematics (STEM), I believe that it is important to understand the kinds of orientations to the future that we are preparing our students for. It will not be conscionable for educators to send students into the world with, for instance, ambitions to control nature and societies for the benefit of a small cadre of people. Secondly, understanding how practitioners of science and technology behave may help educators not misrepresent these practices in the classroom. Besides the obvious job-preparation aspect of getting students prepared for their future careers, mismatches between school and practitioner science and technology can have negative influences on individuals and societies, as when individuals realise that they may be asked in their jobs to do things that they may personally object to (for instance, build weapons that hurt people), or worse, when they attempt to remake society into versions of ideality espoused by well meaning but ultimately mistaken educators. While there will be some degree of overlap, in this chapter I will mostly focus on the macroscopic/historical view of science and technology, and in the next chapter, I will focus on its individualistic practices.

If you recall, an important concept I introduced from the previous chapter was that of technological determinism. Specifically, technological determinism is the idea that inventors work independently of societal norms, following internal technical logic, and/or that when ready, such inventions are thrust upon society which often upset the dominant social order, making changes in the way societies are organised. A dominant line of thinking in sociological studies of science

DOI: 10.4324/9781351116220-5

and technology has been a thorough dismissal of technological determinism. It is neither true that inventions stand apart from societies or that inventions will necessarily cause social change; rather both processes stand in dialectical relation to one another. Inventors often respond to social needs, creating artefacts and systems of human organisation which require more than technical superiority to be adopted widely. Taking the dialectical position in this chapter, I want to argue that there appears to two dominant ways of thinking about the process of invention. Sociologically, the concern is with how innovations are influenced by, and change, societies. On the other hand, psychologists have been interested in the individualistic processes. Social psychology has attempted to bridge the divide, and it is from this discipline that I shall draw some conclusions for makerspaces and science education.

To outline the argument in this chapter, science and technology appear to be predicated upon mechanistic notions of control and subservience. In its worst forms, interpretations of scientific and technological knowledge and know-how have been responsible for many of the worst crimes in humanity, as in the industrialised murder of Jewish people in the Second World War, optimised by the cutting edge computerised record keeping and engineering practices of its time. The question we have to ask is the degree to which such profoundly sociopathic acts have its basis in scientific and technological ways of thinking, or how much of science and technology is amenable to the influence of evil. The conventional thinking may suggest that science and technology are inherently neutral pursuits and that it is the corrupting influence of politics and socialised hatred that compel science and technology into ethical darkness. Yet, a historical study shows that there is at least a case for considering that if full complicity cannot be attributed, the patterns of thought underpinning science and technology make particular forms of evil particularly easy to contemplate. If nature of science (NOS) studies have shown the inseparability of (scientific) knowledge and power, and that what are deemed acceptable scientific problems to pursue are influenced by societal needs, then it stands to reason that society can have evil desires that can be justified and amplified by science and technology. STEM are methods of conceiving and solving problems in the world, and such methods are essentially not unique. We should consider if science and technology, while giving us powerful lenses to view the world, also tint the view in ways that we may not find ethically defensible. As for historical cases which highlight the grand crimes, so too must we consider the mundane, routine, (perhaps more inconvenience rather than) evil that visit us on a daily basis. The question that needs to be asked here is: are there better ways, and if so, how is it that we can educate our students to think of these alternatives?

Before proceeding further, it is important to have some clarity on terminology. Invention, innovation, and creativity have different meanings in different contexts. In this chapter, I want to distinguish these terms as follows: I refer to creativity as the individualistic, psychological activity of coming up with a novel idea. I take innovation to have a social component that includes whether or not the novel idea, artefact, or technology is adopted by a local community or the wider society at large. Finally, I consider the noun form of invention to refer to the novel idea, arterfact, or technology; its verb form will refer to the act of creating especially technological artefacts or systems whose novelty may be meaningfully adopted by groups of people. Inventions are the things that a patent office would judge. So, for instance, a social media network is every bit an invention as a means to detect gravity waves.

How do inventions happen?

If makerspaces are to be sites for the exploration of and instruction in technology, it is incumbent upon us to have a good understanding of the nature of technology. According to Brian Arthur (2009), it would appear that for the most part, we have a fairly good grasp of the working of individual technologies, at least at the conceptual level, for most of the things that we need to live our lives in contemporary times. For instance, even though there may be specific procedures that have been developed to cope with particular contingencies such as material limitations, the working principle of the computer upon which this document is reliant upon is fairly well known. Semiconducting (and other) materials are placed in particular configurations to form logic elements, which process information. Successive layers of translation and decomposition convert human readable text on a screen into binary codes and, ultimately, electromagnetic signals that are reliant upon physical processes that have been harnessed in meaningful ways to human desires. However, while the wide range of individual technologies may be well understood, Arthur (*ibid.*) asserts that the same cannot be said about technology in general. The situation may be compared to our knowledge of the NOS about a century prior—scientific developments, especially in physics, were at its zenith, and public understanding of science was largely celebratory, saluting science as a truth generating device *par excellence*. The parallel for technology is in the way that, especially after the economic success of regions such as Silicon Valley in California, technology is considered as a wealth generating device. To be sure, the relationship between (science and) technology and wealth generation has had a long history. As mentioned in the previous chapter, technological determinism has been a means of understanding the relationship between technology, societies, and the wealth that accrues from its use. Here, we have to be careful to consider technology in its widest sense, to include such things as animal-assisted farm implements such as carts and ploughs and not merely contemporary 'high technology' implements. On the surface, a reasonable (capitalist) explanation still in use suggests that possessing technological tools such as these allowed one greater efficiencies, which therefore made possible the collecting a surplus of some sort (Patel & Moore, 2017). Having such a surplus meant that one could devote more time to the leisurely pursuit of contemplation and hence improvement of further technologies. While this may work as a simple explanation for agrarian and less developed societies, more sophisticated explanations are needed for contemporary times where specialisation has created classes of individuals whose sole task is to actually work on coming up with inventions.

In this regard, the work of Arthur (2009) has been useful in understanding technology and the (sociological) processes of innovation. A key insight has been the use of the theory of biological evolution to explain the growth and change of technologies. To be sure, he does caution that a purely biological evolutionary explanation may not be sufficient: 'The jet engine does not arise from the cumulation of small changes of previous engines favoured by natural selection'. It is not as though minor iterations of jet engines do not develop as component changes are made to lower cost, increase durability, increase efficiency, or any manner of improvements in response to external desires. It is rather more difficult for biological evolution to explain the seemingly spontaneous and inexplicable leaps from, say, shaft-driven aircraft propulsion units to turbine-driven ones. The difficulty arises not from attempting to explain how things evolve, independent of creative agents such as human intelligence, but precisely because it seems like there is an impenetrable barrier when it comes to how human minds engage in creative acts. Here, it is important to make a digression to help ensure that pernicious myths regarding biological evolution are not further spread. Biological evolution can and does explain the gradual accumulation of 'design features' in self-replicating organisms despite the apparent 'strange inversion of reasoning' (Dennett, 2009) that 'in order to make a perfect and beautiful machine, it is not requisite to know how to make it'. While we cannot imagine jet engines spontaneously assembling themselves from a pile of scrap, biological complexity requires no such thing. Through successive iterations of self-replicating organisms whose replication is randomly inaccurate, and whose fitness for the environment is tested over multiple generations, it is possible for beneficial variation to accumulate to give the *appearance* of exquisitely designed mechanism.

The significant difference between biological evolution through natural selection and the invention of design solutions appears to be the distinction between a 'bottom up' and 'top down' design: biological evolution generates solutions to problems of survival and fitness for reproduction from the random, undirected exploration of the solution space. On the other hand, invention can and does involve searches from a top-down analysis of the problem parameters which can reduce the solution search time by pruning out irrelevant branches of the solution space. Arthur refers to this distinction as between a phenomena-based and a needs-based approach. Another significant distinction between biological evolution and the invention of artefacts appears to be between genetic and memetic forms of storage and transfer of information. In biological evolution, information about how to make organisms is stored and reproduced in biomolecules, which necessitate relatively lengthy processes of reading and writing. On the other hand, memetic evolution is communicated via language and sociocultural channels (e.g., gestures, artefacts), the obvious advantage being the rapidity through which information can be spread, manipulated, and transformed. In addition to efficiency over biological means, memetic evolution through language has the advantage of being the sole 'digitized basis for reliable cumulative evolution':

> (It is digitized in the sense that it is composed of a finite set of discrete, all-or-nothing elements—phonemes—that can survive noisy transmission, different accents and tones of voice, drawls and lisps, by a process of largely automatic correction to norms.) Other species, such as chimpanzees, have a handful of culturally transmitted traditions—of termite fishing or grooming signals or nut cracking, for instance—but nothing that ramifies the way human culture does. Language, by providing a basic repertoire of readily replicated elements, permits the reliable transmission of semiunderstood formulas, recipes, admonitions, techniques. (It is not typically noticed that one of the most valuable features of language is its ability to convey information down a chain of communicators who do not really understand what they are "parrotting.")
>
> *(Dennett, 2009, p. 10064)*

Language serves not only to ensure rapid communication, it provides a means to reliably send meaning down channels that may be noisy. If invention, at least at the social, level can be interpreted as a form of memetic evolution, then we begin to have some insights into how inventions spread and are modified by communities. Arthur (*ibid.*) suggests that inventions are composed by the combination of modularised ideas. For instance, as the motor vehicle is made up of an engine (electric, internal combustion, external combustion, biological 'combustion') connected to a chassis supported by wheels. There are standard problems that have been repeatedly encountered across different contexts, which quickly acquire solution sets that describe the multiple standard solutions that are possible for the problem. These solutions then become incorporated into the standard repertoire, ready to be combined into subsequent solutions. Similar to evolutionary spread of mechanisms for survival, these standard solutions can then circulate among problem solvers as literal tools when made into artefact form or as thinking tools (Dennett, 2013) when used in its metaphorical form.

Significantly, Arthur (*ibid.*) reminds us that these solution units do not necessarily represent the best solution possible. These ideas exist in the social space and do not always get tested against physical criteria of 'fitness' for purpose. As a result, what gets selected to become the 'go-to' solution for particular problems may merely be good enough and be prevalent because of sociocultural reasons rather than technical superiority. In the last few decades, examples of technically inferior designs that nonetheless gained widespread adoption would include the VHS video system winning out over the superior quality Betamax; Digital Audio Tapes; and the Concorde supersonic passenger aircraft. In all these cases, technical superiority did not help its widespread adoption, and crucially, factors other than technical superiority were significant in making these decisions. If we recall the distinction between the arbitrary and the obligatory in considering the social nature of truth claims, there are important lessons here for education. In deciding the curriculum for makerspaces, a problem arises as to whether instruction should focus on technologies that are technically superior or on the social circumstances that lead to the adoption of inferior technologies. Certainly, as should be clear from the focus of this book to this point, my position is that both positions deserve attention in schools. Students should understand what counts as a technically superior solution, because these criteria form the basis of what is *true*. On the other hand, social conventions dictate what is *correct*. Students need to know the difference between the two, and how we might come to judge solutions to be correct for the social purposes that we may or may not explicitly announce as desirable. For instance, while the Concorde flew faster than any commercial airliner, the high price of its tickets and its sonic boom made the aircraft unpopular among potential passengers and people who lived under its flight path.

Arthur (*ibid.*) also reminds us that technological innovations are essentially 'a phenomenon captured and put to use'. Its success can derive either from a genuinely useful phenomenon that eventually finds a niche for its commercial success or from its developers responding to an existing need in society. For instance, the laser was first theorised as a consequence of quantised behaviour of atomic energy levels, and the existence of a means to produced laser light was predicted even before people knew what to do with it. Conversely, in aircraft propulsion, initial designs made use of piston and propellor designs, which immediately encountered problems as aircraft flew higher and faster. There was a need to address the reduced oxygen at higher altitudes, and the increased drag that the propellors presented at higher speeds. In the case of the laser, one could say that it had been a solution looking for a problem, whereas the converse was true for aircraft propulsion. In either case, the concept of '*capturing* a phenomenon' and 'putting it to *use*' deserve our careful attention here. I want to argue in the next section that the possibly exploitative tone may not entirely be accidental. Because innovations are social things, in that the design responds to a need that is felt by a large enough segment of society, knowledge of innovation encapsulates some form of knowledge about society, to a far greater extent than knowledge in the natural sciences. In other words, it is not enough in learning to be an engineer or designer to only have proficiency in its technical aspects. Knowledge of the social relationships within communities of technologists, and between technologists and the wider society at large, is also important for students to acquire. Knowledge of the former provides students the necessary tacit knowledge to negotiate the boundaries of practice that students need to work within; knowledge of the latter helps students avoid making errors in judgment when deciding what ought to be made. In the following section, I will review what may be some excesses of science and technology, with a focus on some of its underlying philosophical motivations. Science and technology, I want to argue, may not be the neutral knowledge that is abused in societies by nefarious elements demanding scientists and technologists to do evil. Instead, there may be elements within these disciplines that make wrongdoing especially likely.

The dark side of innovation

A curious observation was made since the 1980s that there appears to be a relationship between science and technology (specifically, engineering) and radical Islam. More than any other group of university trained graduate, engineers, by far, are overrepresented among violent jihadis. It is not as though an engineering training necessarily results in its graduates becoming susceptible to violent ideologies; the overall number of engineering trained violent jihadis as a fraction of the total number of engineers is small. However, in comparing educational qualifications among extremists, engineers appear to be over-represented. Gambetta and Hertog (2009, 2016) studied a sample of 404 violent jihadis over seven major data sources, ranging from the team responsible for the 1993 World Trade Centre bombings in the United States, the September 11 suicide attacks, information on the Jamaa Islamiya terrorist cells in South East Asia, and other information from the Middle East and around the world. Of these 404 individuals, 284 had known educational information; 196 of these had higher education, either complete or incomplete. Of these 196 with higher education, they could find the educational specialisation of 178:

> the group that comes first by far are indeed the engineers: 78 out of 178 individuals had studied this subject. This means that 44% of those whose type of degree we know were engineers. On the whole, the individuals who studied for what we may call "elite degrees"—engineering, medicine, and science, generally the most selective programs in the Islamic world— represent 56.7%. (p. 204)

In comparison, the population average of engineers across the countries of origin of these extremists, weighted to represent the sample distribution, is only 3.5%. As an interesting aside if only because I call it home, leaving Singapore out of the calculation brings the 'background rate' of engineers across these countries of origin down to 2.1%; Gambetta and Hertog (2009) note that Singapore has 'an extraordinarily high number of engineers'. Even if the 78 engineers are compared against the entire sample size of 404 individuals, this makes the percentage of engineers within the militant group 19%, a much larger number than the proportion of engineers within national populations. An important caveat of this study is that violent extremism studied here does not appear to stem solely from their academic training alone. Gambetta and Hertog (*ibid.*) also compared violent extremists in other parts of the world, across different time periods, and across different parts of the political spectrum. While engineers were present, in no other case than in violent Islamic extremism were they overrepresented to such a degree.

Candidate explanations for this state of affairs included network diffusion effects, and selection based on technical ability. Network diffusion is the explanation that the initial members of a group are likely to recruit for other members

of similar traits. When examined, Gambetta and Hertog find that, across uncon-
nected terrorist groups, the preponderance of engineers continues to exist. Even
within independent organisations where the size of the engineering segment
could be explained by network effects, what cannot be explained are 'both the
disproportionate share of engineers who became prime movers and their greater
willingness to stay in or join a radical network even if started by non-engineers'
(p. 214). As for technical capability, it turns out that for most of the violent
attacks, the technologies required to be deployed were in fact rather simple. In
organisations like Hamas, engineers are in senior managerial positions distant
from the technical responsibilities. Significantly, the presence of engineering
talent did not correlate with an increase in destructive ability, as when several
separatist movements with devastating attacks were composed of poorly educated
and working class individuals.

Gambetta and Hertog (2009) assert that since we can rule out these explanations, what remains to be the key factor in understanding this problem may
be the engineering mindset. Gambetta and Hertog suggest that the engineering mindset hypothesis should present itself in two related findings: 'first, we
should find engineers to have a greater predilection towards joining radical
political groups regardless of Islam […] Second, engineers compared with people in other disciplines should also manifest more radical views. This prediction
too should be verifiable independently of Islamism' (p. 215). While the rest of
their paper and their entire book makes for a fascinating read, the authors do
make a careful and convincing case in support of these predictions. They find
in extremist Islam an 'ideological cocktail' of a corporatist and mechanistic
worldview of an ideal society. 'Extremist Islamist literature rejects Western plu-
ralism and argues for a unified, ordered society ruled by a strong Islamic leader,
in which an unassailable division of labour is created between men and women,
Muslims and non-Muslims, political leaders and their flock. The fear of social
chaos undermining this established order is a leitmotif of Islamist thought'
(p. 220). Engineering and the natural sciences seem to be ideal candidates either
to nurture minds, or attract minds, that are particularly susceptible to cognitive
closure and the existence of clear-cut, sharply delimited binarisms of right and
wrong, as opposed to forms of knowing such as the humanities which promote
a more tolerant acceptance of ambiguity. Gambetta and Hertog (*ibid.*) go on to
cite philosopher Friedrich von Hayek, who observed in 1952 that engineering
fosters in its students a view towards rationalised control of processes plays a key
role. Instead of 'understanding individuals and their world as the outcome of a
social process in which spontaneous behaviours and interactions play a significant part', engineers are more likely to insist that 'societies operate in an orderly
way akin to well functioning machines' (p. 222). Rather chillingly, von Hayek
suggested that:

> It is not surprising that many of the more active minds among those so
> trained sooner or later react violently against the deficiencies of their

education and develop a passion for imposing on society the order which they are unable to detect by the means with which they are familiar.

(Hayek, 1952, p. 102, in Gambetta & Hertog, 2009)

Gambetta and Hertog do point out that the conditions of relative deprivation experienced by these extremists appear to be the trigger that sets the conflagration alight. Compared to engineers in other countries who manage to lead successful and fulfilling lives as valued members of society, many of these violent jihadis have had the experience of alienation, often as a consequence of studying abroad in elite institutions, in a foreign culture that contradicts their early upbringing. These experiences of relative deprivation and alienation combine with political repression and resentment in countries where the only possible reaction is either acquiescence or violent rebellion.

To some degree, my inclusion of the 'Engineers of Jihad' hypothesis (derived from the book of the same name by Gambetta and Hertog, 2016) can plausibly be decried as borderline alarmist. My inclusion into this book is not meant to imply that a training in engineering will necessarily lead students down the road to violent extremism. It is important to understand what is claimed here by Gambetta and Hertog. They do not claim that the engineering mindset guarantees that one is especially prone to violent means to achieve one's ends. However, they do claim that the engineering mindset is particularly compatible with an attitude that seeks distinct boundaries and clean-cut categorisations, is intolerant of ambiguity, and significantly, gives people the tools that they need to assert these particular visions of utopian order onto the external world around them. More than the social sciences and the humanities which deal with fluid categories and multiple perspectives to thinking about phenomena, it is engineering that deals with the technical inevitabilities, with a concern about the means to attain idealised ends. Considering that the engineering discipline derives its historical roots from the construction of siege engines used for tearing down what was built by the architects, engineering is the quintessential 'war built' discipline (Nieusma & Blue, 2012). Even outside of war fighting organisations, engineers can be peculiarly prone to hierarchical authority structures; the question that is often asked is whether engineers work best as employees of organisations, merely carrying out the desires of management, or if they should have their own professional associations that can assert some form of oversight over the work of other engineers (*ibid.*).

It should be obvious that the social and ethical dimension of engineering needs attention. Philosopher Bernard Stiegler (Le Média, 2018) has raised concern over the veritable hostage situation of the manner in which we as a human species are reliant on technologies. Stiegler recalls the concept of *exosomatisation* first used by Marx and Engels to describe the human condition: among all species, it appears that only humans are reliant to such an extent to organs that are not organic to our bodies. For instance, our abilities to live outside of the equatorial zone, in regions where winters would kill us or require continual migration, is dependent on warm clothes, shelter, and heating, none of which are innate. Continuing the march of 'progress', we come to our contemporary times, where many would make light of the situation that internet access is akin to oxygen. The bottom line here is that we require expertise on how to produce these veritable organs that ensure our survival. Access to tools and technologies becomes a matter of social domination and control: who has it and who determines what these tools should be used for are questions that are usually not part of the core engineering curricula, because these are conventionally thought to be 'not engineering concerns'. This should raise many red flags: in providing students with privileged information that holds many lives hostage, surely we want to be sure that individuals with malfeasant intentions are not able to affect their goals. One could complain that there is no easy way of stopping such bad actors: anyone with access to a knife or a motorised vehicle could already do damage. While this may be true, contemporary technologies literally make facile the efficient ending of many more lives. Ravikant (Reid, 2019) reminds us that control of passenger aircraft was ceded to one such group of bad actors, dramatically scaling up the available destruction, and its inevitable overreaction. More worryingly, Ravikant points to existential threats such as synthetic biology: with access to gene editing tools such as CRISPR now available to even well-equipped high-school laboratories, it is possible to trivial to develop weaponisable strains of lethal viruses such as H5N1. In addition to the ubiquity of the tools that can be used for both good and evil, there is a statistical asymmetry to worry about here: there are simply more ways for take apart, destroy, sabotage, or otherwise render unusable an artefact or technology, than there are ways that for such a device or system to be functional. While knowledge of how technologies work can enable its pro-social deployments, it can also amplify socially questionable and even downright sociopathic intents. With the proliferation of public informational resources, the argument of security through obscurity cannot stand. In addition, many of these dangerous technologies have dual or multiple uses, and access to these tools cannot be meaningfully restricted.

That the guilds, associations, and educators, as keepers and promulgators of this knowledge, have an ethical obligation should be very clear. Educators in general have always been tasked with the nurturance of pro-social orientations in our young: we teach them how to use scissors and that they should not cut others with it. Yet, I find it incongruous that we teach older versions of these same children about robotics and automation (for instance) often in a celebratory manner of the bright shiny futures they represent and neglect to tell them anything about the socio-economic implications of these technologies. It has been that the role of educators cannot be limited to merely transmission of conceptual knowledge; the challenge for us as STEM educators here is the communication of the contentious 'non-STEM' knowledge of ethical uses of our knowledge. I want to argue here that beyond the explicit communication of ethical knowledge lies the tacit knowledge associated with curriculum and pedagogical decisions such as whether or not the ethical dimension of inventing is even part of STEM programmes, and the manner which teachers convey what STEM is about. It cannot be enough for teachers to append an 'ethics module' after the fact, suggesting that the core of STEM knowledge has nothing to do with considerations of the ideal deployment of technical knowledge. The manner in which we discuss STEM knowledge should also be considered. While economistic considerations at least implicitly convey a celebratory vision of how the future will be improved with new technologies (and therefore that students should desire to be part of the class of recipients of this largesse), teachers can and often adopt an indifferent attitude to conveying STEM knowledge, preferring to focus on the technical details. The possibility exists however in communicating STEM knowledge, there is no such thing as a disinterested position. The disciplinary specialisation of STEM, specifically, the ways in which problems are framed and perceived and solved, makes particular solutions more likely. It is as the popular aphorism has it: if the only tool one has is a hammer, all problems start to look like nails.

The problem with science

So far, we have seen that the engineering perspective makes one more prone to accept rigid demarcations between categories, and to enforce these demarcations as a practice of one's craft, to the extent of using violent means against other human beings when one perceives the helplessness of one's position. We have also seen in previous chapters that the scientific orientation towards the generation of truth claims can cause problems, as when, for instance, social researchers exhibit 'physics envy' and attempt to adapt scientific methods to make inappropriate claims. Most visibly, these debates have centred around the use of quantitative methods and a general antipathy towards qualitative methods; these methodological 'wars' have resulted in books on research methods to preface themselves with an introduction to the epistemological value of these methods (e.g., Denzin & Lincoln, 2000, 2018). Indeed, there has been significant contention over the relative 'scientific merit' of using different methods in education research, with researchers arguing for the unique degree of complexity of educational phenomena (Labaree, 2004; Luke, 2011; Wieman, 2014). While I will return to these matters soon, the pertinent question at this point is, again, the degree to which the scientific perspective may offer us biased ways of perceiving the world, interpreting phenomena, and deciding on courses of action. I believe that reality does not 'speak for itself', rather that human beings interpret meanings and place value onto what occurs. For instance, there is nothing inherently *true* about interpreting lightning as a stream of ionised plasma formed by the breakdown of air molecules as dielectric media separating charged clouds from the ground versus older mythologies that lightning is the work of gods. The modern explanations are more *correct*, in that more accurate predictions can be made with these ideas, but as with the ubiquitous 'guess what's in the black box' type of activity that we use to introduce the 'scientific method' to students, there is simply no way for humans to ever open up the box to know if what we suppose is true. We make simplifying assumptions all the time, value particular forms of explanation (natural versus *super*natural), and still invent imaginary particles and interactions to explain 'what is in the box'. While we have largely been spectacularly successful in these ventures, it is important to consider what these means of simplifying the complexity of reality have contributed to our ways of thinking about the world and acting in it.

In this regard, it is important to take stock of the uses and abuses of science and technology. There have been multiple ways that our lives have been transformed by science and technology over the years since we have had its modern interpretation associated with the scientific revolution in Europe in the mid-15th century. Life expectancies have risen, and infant mortality has fallen, such that there are more humans on earth now than ever before. At the same time, we are regularly confronted with anxiety causing messages about how our collective humanity is destroying the very means of living, while for the most part, little can be done to change things. The situation is complex: we should avoid the technological determinist position blaming scientists and technologists for coming up with these technologies. Yet, if we are to have a hope of a different kind of future, we should at least have a better sense of the kind of thinking that has gotten us to where we are and attempt to educate in a manner that can give us a qualitatively different way of thinking about phenomena. In the last chapter, we have seen the trouble that a reductionist viewpoint has gotten us. Tightly associated with this has to be the degree of control over nature that is connoted and expected by the scientific enterprise. Scientific knowledge has brought us great control over our natural world. We no longer occupy 'demon haunted worlds' (Sagan, 1995); things do not happen because mystical entities will it so. While this perspective has been successful, it is reliant on the notion of a stable, predictable universe and has a tendency to overstate our confidence in the scientific knowledge and principles. This overconfidence is expressed in such aphorisms as Einstein's 'god does not play dice' or, most controversially, in the writings of Francis Bacon:

> For you have but to hound nature in her wanderings, and you will be able when you like to lead and drive her afterwards to the same place again. Neither ought a man to make scruple of entering and penetrating into those holes and corners when the inquisition of truth is his whole object.
>
> *(Soble, 1995)*

Such a position has been criticised by feminist philosophers of science such as Sandra Harding (1991), as implying the dominant philosophy of science to portraying or even advocating a metaphor of rape. To be sure, the feminist position is not without critics. For instance, Soble (1995) defends Bacon, proposing that Harding's definition of rape may be too encompassing, in that rape happens *any time* a woman does not initiate the act. While it is not my intention to wade too deeply into the waters of feminist critiques of science, I believe that it is sufficient to state here that there is at least some contention that the metaphors of scientific process include at least some component of coerciveness. Depending on one's politics, such a move could be interpreted from a range that includes persuasion, coaxing, through towards torture to outright rape. What is less contentious is the notion that once nature gives up its secrets, it is as if the game is up, and we have everlasting understanding of the way that nature functions. Such a position encourages an approach to science education that privileges the communication of the known relationships between properties of nature that we have arbitrarily determined to be significant.

While I will not go so far as to claim that the teaching of science makes one particularly prone to violence to achieve one's means, the approach to thinking about scientific knowledge as final and absolute surely is not helpful to science educators. Scientific knowledge progresses, but at the level of schooling, such changes are not obvious. Couple this observation with school systems where accountability pressures are high, and where teachers' classroom authority can be strongly based on their being the epistemic authority in the class, the temptation to focus strongly on the knowledge products of science can be high indeed. Typically, teachers start students off with science in primary school seemingly as a form of magic, where we are able to predict counter-intuitive occurrences, and then we proceed to middle school where experiments that do not work as predicted are dismissed as 'experimental/human errors'. Eventually at the undergraduate and later levels, scientists need a different orientation to knowledge and the means through which we obtain it, often focussing deeply on the 'errors' of prediction. It is almost as if we teach students successively more truthful approximations of our understanding of nature. Take the models of the atom for instance: because of perceived difficulties either in teaching or learning the material, we present models of the atom as billiard balls, planets, standing waves, before telling them about probability distribution clouds. While it would seem that the training of scientists may not (?) be hampered by such a process of successive 'lying' that approximates the truth, the consequences for public understanding of science can be more severe.

As Stuart Firestein (2012) has surmised in his book, much of the scientific pursuit is about ignorance, in that the usefulness of scientific knowledge is only in allowing us to formulate better questions to further interrogate (pardon the metaphor) or inquire about nature; these pieces of knowledge are important not only because they allow us some degree of control over our world, but also (for scientists in particular) they serve as stepping stones. The concern here is with the public understanding of science: for students who never make it to a career in science, and who only have passing knowledge of science, the impression that these people (who will form the vast majority of most populations) will have is one of science as, essentially, understood magic tricks; its practitioners magicians. Firestein gives a particularly striking example of neurologists studying electrical activity associated with the brain. Called 'spikes' for the way they appeared on traces of paper, Firestein reports that:

> For the last 75 years my neuroscience colleagues and I have been studying spikes and teaching our students about spikes and making grand theories about how the brain works based on spiking behavior. Some of it is true. But what have we missed by concentrating on spikes for the last eight decades? A lot, it turns out. There are many other electrical sorts of signals in the brain, not as prominent as spikes, but that's a reflection of our recording technology not of the brain itself. These other processes, as well as chemical events that are not electrical, and therefore can't even be seen with an electrical apparatus, are now being recognized as perhaps the more salient features of brain activity. But we have been mesmerized by spikes, and the rest has been virtually invisible, even though it is right in front of our faces, happening all the time in our brains. Spike analysis was a successful industry in neuroscience that occupied us for the better part of a century and filled journals and textbooks with mountains of data and facts. But it may have been too much of a good thing. We should also have been looking at what they didn't tell us about the brain. (p. 26)

Having knowledge is useful, but it can also handicap us from other ways of thinking about phenomena, which may prevent us from accessing different kinds of insights. In instructing students, the temptation for science teachers is to occupy a position of authority, precisely because there is an informational asymmetry between teacher and student. Given that, for the most part, scientific knowledge enables a certain degree of control and prediction, the public perception of science and technology that is formed by the interaction of novices with well meaning science educators will certainly tend towards the impression that we have dominion over nature. In order to convince students of the correctness of the laws and principles that they will instruct, it is fairly common practice to perform demonstrations. Indeed, a common instructional procedure for science is the Predict-Observe-Explain (Liew & Treagust, 1995) protocol, where students are given a physical scenario to which they are invited to make predictions as to what may happen next. Typically, these predictions will be contradicted by the physical system, leading to a state of cognitive dissonance whereby teachers insert themselves as knowledgeable experts to resolve the dissonance. Students quickly learn that it is a good idea to learn science, because of its ability to predict and control nature, and teachers quickly gain an aura of respectability accorded to minor priests in religious orders.

I am only being half facetious in comparing science teachers to priests; in an essay entitled *How to defend society against science*, Paul Feyerabend (1975) objects to how:

> Scientific 'facts' are taught at a very early age and in the very same manner in which religious 'facts' were taught only a century ago. There is no

attempt to waken the critical abilities of the pupil so that he may be able to see things in perspective. At the universities the situation is even worse, for indoctrination is here carried out in a much more systematic manner. Criticism is not entirely absent. Society, for example, and its institutions, are criticised most severely and often most unfairly and this already at the elementary school level. But science is excepted from the criticism. In society at large the judgement of the scientist is received with the same reverence as the judgement of bishops and cardinals was accepted not too long ago. (p. 4)

To be fair, the historical context of the time when this was written is important, as I discussed earlier, the period when Feyerabend was writing has been associated with the height of exuberance of the movement sceptical of scientific claims. There is almost a certain amount of glee, as if Feyerabend and others were pointing out the nakedness of the emperor. With some distance, it is now certainly clear that these arguments may have been in excess. However, the sentiment contained in these critiques is not essentially wrong. As I will detail in the next chapter, the practice of science is not, and should not be, treated as a guaranteed truth generating mechanism. Even at its very best, the relationship between empirical evidence and the representations that we use to express the truth claims are not always free of interpretation, human agency, and the consequent subjectivity. Yes, there are truth claims which are more or less true, and we can, as a society, tell them apart. Yes, many scientific claims tend to lie on the side of greater truth value. However, the risk exists that we abuse the goodwill that society places on the certitude that the scientific enterprise appears able to deliver:

> In conclusion it is, perhaps, desirable to remind the reader once more that all we have said here is directed solely against a misuse of Science, not against the scientist in the special field where he is competent, but against the application of his mental habits in fields where he is not competent. There is no conflict between our conclusions and those of legitimate science. The main lesson at which we have arrived is indeed the same as that which one of the acutest students of scientific method has drawn from a survey of all fields of knowledge: it is that "the great lesson of humility which science teaches us, that we can never be omnipotent or omniscient, is the same as that of all great religions: man is not and never will be the god before whom he must bow down."
>
> *(Hayek, 1952, p. 102).*

With most conventional approaches to science education, the material is presented in an uncomplicated manner; much of science education research is interested in more and better methods for the effective communication of particular difficult concepts as if the educative problem were only one of communication.

The actual problem, recounting Firestein's notion of the significance of igno-rance to science, is communicating our uncertainty and the vast amount of igno-rance we have in comparison to what we already know. As Hayek recommends, and I ask of science and/or contemporary STEM education: when is it that we emphasise to students that 'we can never be omnipotent or omniscient'?

It would appear that these historical critiques have not lost its currency in the intervening years. Indeed, widespread interest in 'the nature of science' as a goal for science education has only taken place since approximately the turn of the 21st century. For Collins and Evans (2017), their survey of the public uses and abuses of science has presented such a dire picture that they believe that a book length treatment was desired. To be careful, the thrust of their work is that public confidence in settled arguments in science requires defending from particular political actors. These seek to capitalise on widely misunderstood notions such as the false equivalence between the scientific position and any of its widely circulated antitheses. In any case, the implications for science educators are clear: we have widespread misunderstanding of science, tending towards the polar extremes of excessive mistrust or over-exuberant faith. We distrust science when we fall for charlatans who capitalise on doubt which is part of the scientific process, and we overstate scientific certainty when we cast human sociopolitical decision-making in scientific rational terms. As with most things, balancing on the tightrope between two contradictory tensions can be the hardest thing to do; when instructing students, it is usually the easiest to fall securely on either side. Yet, the challenge is to demonstrate the tension that exists, to convince students that certain truth claims in science are settled, and yet many others are not. In other words, an ideal science education should, as Firestein suggests, present us with insight into where the boundaries of our knowledge end, and where igno-rance begins, and how these regions are related.

Why makerspace?

It is into this gap that I believe makerspaces should fit into. Conventionally, makerspaces have been portrayed as merely engaging spaces as an instructional tool to enhance student engagement through the use of novel tools and tech-nologies. For me, the value of such a space lies with the interaction of humans, natural phenomena, and the manner in which we may derive partial truth claims about such phenomena. I take it as a given that our senses, instruments, and the general method of the sciences which privileges reductive approaches, our rep-resentations of nature reveal only very partial understandings. These are doubt-less useful, but as Feyerabend suggests, may merely be our 'best worst theories'. Instruction in science should communicate such aspects of science education. Specifically, students need to know how it is that scientists are able to make these claims, given that, 'mother nature' does not give up its secrets readily. What are the physical processes that scientists have decided that matter? Why have scientists decided to reduce the complexity inherent in nature to merely the set of terms currently accepted? While we conventionally accept that there are 'experimental errors' that limit the precision of our readings, what 'lies beneath/ beyond' if we were to investigate and demand, for instance, an additional significant figure's worth of precision? In other words, in addition to the end products of knowledge creation by scientists, students should become competent in the processes of knowledge creation too. It is to these ends that most practical investigations and demonstrations have been directed towards to date, but the argument here is that contemporary sciences as practiced in working laboratories have proceeded apace, and make use of a wide array of technological implements, many of which are commodity items, or have equivalents well within the reach of almost everybody.

Speaking about the physical sciences, for instance, computational techniques, and 'mechatronic' integration of sensors, actuators and microcontrollers have largely supplanted raw visual senses in measuring physical variables. Data loggers have served this function in many school laboratories and are admirable for their ease of use and easy integration with many possible lesson formats. However, in order for these devices to be 'user friendly', a certain degree of black-boxing is often done to hide the internal functioning of these devices from its users. It is certainly contentious to suggest here that opening up these black boxes can afford learners particular outcomes, especially since the added complication can easily hinder the acquisition of what is traditionally deemed to be the more important knowledge of the scientific principles. What it comes down to, I would argue, is a question of the relative valuing of product over process. To replace data logging products with 'off the shelf' components, the end result may be an experimental apparatus that can be especially prone to error and malfunction, but such a set-up may precisely be the point: as I will develop in the next chapter, it appears that for the most part, scientific claims are the result of experimental conditions where nature is carefully coaxed into revealing particular chosen facets. The process is tentative, and the set-up we use for amplifying our senses requires a constant 'fiddling with' in order that we can get the results that we are looking for.

Besides the notion that scientific truth claims are based on empirical processes that are tentative and difficult to orchestrate, there is also the aspect of the authenticity of the types of science that we communicate to our students. The forms that science has taken in schools over the years seem to have frozen in place since about a century and a half ago (Gunstone, 2013). Specifically, we conventionally teach science in the siloed form of physics, chemistry, and biology, conveniently ignoring, for instance, inter- and trans- disciplinary approaches such as biochemistry, biophysics, and the general form of computational methods that have come to influence the ways in which the sciences are practiced and even thought about. For instance, chaotic systems such as fluid flows are effectively impossible to solve analytically. However, good science is still being carried out by modelling such systems '*in silico*' on computer simulations.

To summarise this section then, I consider makerspaces to be a possible solution for an over-deterministic understanding of scientific practice and an over-confidence in our scientific knowledge. Additionally, if students should learn how scientific knowledge is created, at least some of their experience with science instruction should involve the messy contingencies involved in coaxing nature and our representations to coincide. In the next section, I wish to discuss the matter of creation, more generally speaking to include both new knowledge, and artefacts.

How does creation happen?

In this section, I wish to discuss creativity, not from a psychological perspective in thinking about what happens in the mind as one arrives at useful novelty, but rather from a philosophical perspective in thinking about the antecedent conditions and sociocultural environment that make creativity particularly possible. The rationale for such an orientation comes in recognition of the social nature of the human species: we are deeply social creatures, and while individual cognition is important, it is more likely that in order to achieve significant projects, cooperation with others is important. In thinking about innovation in science and technology, I assert that it is important to think past the individual account of creation, to consider how human communities and societies come together for the purpose of creating something anew. Especially for the context of education, understanding how creation happens can help educators make particular creative outcomes more or less likely. A concern for educators is that we should instruct students not only in the accepted knowledge bases of the discipline, but we should also make it likely that they will have the abilities to exceed our current understanding and create new forms of knowledge. As I have surveyed above, the new knowledge of science and the artefacts of technology may not necessarily be socially beneficial, and scientists and technologists have no morally defensible position from which to absolve themselves from the responsibilities of their creation. In this section, then, I want to ask: what should idea instruction in makerspaces and STEM classrooms look like, if we wish to attend to these goals for schooling, above and beyond merely creating engaging contexts for the acquisition of STEM knowledge.

Insights from considering a social orientation towards understanding cognition have driven shifts in thinking about cognition in the last several decades, from the notion of distributed cognition (Hutchins, 1995, 2014) to the embodied cognition perspective I surveyed in an earlier chapter. Distributed cognition views coordination of information flows across multiple individuals, communicated through verbal and non-verbal means, as a significant and under-studied perspective in thinking about cognition. Embodied cognition, in brief, considers the body and its immediate environment as a resource for cognition, best encapsulated by the analogy of the circulatory system: while the heart is the major organ of the circulatory system, circulation does not happen without the blood vessels. Similarly, the brain is the main seat of cognition, but it is not complete without the body and the environment that it exists in. Along with these perspectives, the social psychological position for creativity is represented by Teresa Amabile's componential theory of creativity (Amabile, 1983, 2012; Conti, Coon, & Amabile, 1996). In this approach, creativity is influenced by four factors: (i) domain-specific knowledge: one cannot be usefully creative in a domain not of one's expertise; (ii) domain general knowledge: without skills for creativity such as known procedures for divergent idea generation and critical thinking, being creative becomes a bigger challenge than it needs to be; (iii) motivation: while certain forms of extrinsic motivation can stimulate creativity, creativity is generally and more reliably a result of intrinsic motivation; (iv) sociocultural contexts: certain cultural factors, such as the tolerance or even celebration of the inevitable failures that result from creative ventures, can encourage individuals to attempt creativity. On the other hand, societies that discourage, or even punish, individuals for transgressions from established norms tend to be rewarded with conservatism and a lack of innovativeness.

What might be an appropriate orientation towards creativity, and how would a classroom that seeks to nurture creative dispositions look like? Gert Biesta (2016), discussing the perspective from a theoretical consideration of the ideals of education, suggests that a way to think through this problem lies with a consideration of what schools have been increasingly been asked to do. Schools, caught up in the drive to become administratively accountable for the funding that they receive, are no longer trusted to perform their task out of professional discretion. With quantitative measures on the one hand, and scientist research on the other, the demand for schools has been to better guarantee results. In this way, makerspaces and STEM education have been seen as rather obvious priority areas because of their apparent immediate relation to economic opportunities in organisations creating technological innovations. For Biesta, the appropriate orientation to education, seen as an act of creation of particular kinds of individuals, embraces the risk inherent in 'making' individuals. Humans are not machines and the interaction between teacher and student in the process of education is not one of installing an appropriate software update, or bolting on parts on a production line; even metaphorical usage of these ideas are just plain wrong. As Labaree (2004) contends, the educational profession faces a unique challenge in that in contrast to other professions, the cooperation of our clients is paramount to our success. Doctors, engineers, and lawyers (for instance) can do their work without their clients' participation in the process; indeed in many cases for these professionals, clients' participation can be detrimental to the success of their work—surgery, for instance, works best in sedated patients. In education, students can and often do refuse instruction for a variety of reasons. Though we may colloquially refer to these as student aptitudes, the situation is certainly more complex and teacher attitudes towards the nature of this interaction can have a significant influence on the outcome.

Specifically, Biesta (2004, 2013) laments that the discourse around education has shifted towards *learning* and away from teaching. Such discourse has made it more likely to think about education as a simple task of provision of knowledge, and most egregiously, deletes three vital terms that, according to Biesta, needs to accompany the concept of learning. Thinking about learning is not complete without thinking about *what* it is that is supposed to learnt, *why* such learning should take place, and *what is the ideal relationship* between the learner and the instructor. The first two terms are curriculum questions that, for makerspaces and STEM education, I develop throughout this book. As for the nature of the ideal relationship, the focus on learning, especially with the aid of educational technologies, has led to a diminishment of teachers and the role that they play in creating the circumstances more or less likely for learning to happen. In opening his book, Biesta writes poetically that:

> This book is about what many teachers know but are increasingly being prevented from talking about: that education always involves a risk. The risk is not that teachers might fail because they are not sufficiently quali- fied. The risk is not that education might fail because it is not sufficiently based on scientific evidence. The risk is not that students might fail because they are not working hard enough or are lacking motivation. The risk is there because, as W. B. Yeats has put it, education is not about filling a bucket but about lighting a fire. The risk is there because education is not an interaction between robots but an encounter between human beings. The risk is there because students are not to be seen as objects to be molded and disciplined, but as subjects of action and responsibility. Yes, we do edu- cate because we want results and because we want our students to learn and achieve. But that does not mean that an educational technology, that is, a situation in which there is a perfect match between "input" and "output," is either possible or desirable. And the reason for this lies in the simple fact that if we take the risk out of education, there is a real chance that we take out education altogether. (p. 1)

For Biesta, a way to think through this problem of the ideal relationship for instruction is to consider other archetypal relationships of power difference and creation, the analogy being that education should be thought of as an act of crea- tion of particular types of persons. Among creation mythologies, Biesta suggests that the biblical accounts of creation offer strong influence to English speaking communities. At the heart of the problem is the distinction between creation as an act of metaphysics, as *creatio ex nihilo*, versus creation as an existentialist phe- nomena, as the outcome of a series of encounters and events. There is a difference between two accounts of creation, attributed to Elohim and YHWH (Yahweh). The creation of Elohim is one of calling into life by a 'calm, distant, celes- tial, hands-off creator' (p. 14). On the other hand, Yahweh creates as a 'nervous

[…] hands-on micro-manager' (p. 14). Biesta (*ibid.*) summarises the theological argument:

> Yahweh does not so much give Adam and Eve life as he gives them a test of life. "He gives them life on a kind of conditional trial loan to see if they are going to abuse it and try to become like him, in which case he is prepared to withdraw from the deal and wipe—or wash—them out" (Caputo, 2006); this is unlike the story of Elohim where life is what Derrida (1992) would refer to as an unconditional gift. Yahweh, as Caputo puts it, "seems to have a bit of a short fuse, seems inordinately suspicious of his own creation, and is far too nervous about his offspring for a good parent".
>
> *(ibid., p. 69)*

Throughout his book, Biesta makes a convincing case for the ideal interaction in the creation of individuals not in strong, mechanistic terms even though it would appear that administrators would prefer such an orientation. For Biesta, the point of this theological excursion is to provide a philosophical insight into the *quality* of an ideal educational interaction, given that education can be seen essentially as the *creation* of particular kinds of individuals. In this perspective, Biesta's intent is for us to be more accepting of the messiness, dissent, noise, and all manner of associated risks that need to be considered as an attendant part of the process of education *as creation*.

This position is decidedly a different perspective than the current thinking about the manner in which education is to be carried out. According to Biesta, constructivism as a theory of learning seems to have had an outsize influence in this state of affairs. Constructivism appears to have shifted the discourse in education from a focus on the teacher towards the learner. Instead of the artistry involved in communicating knowledge and creating the circumstances that may make learning more or less likely, constructivism encourages a certain mechanistic way of thinking about learning. For others like Roth (2015), if knowledge is constructed in the mind of the learner, it can be extremely seductive to think about the process of education in building terms. Indeed, even the metaphor of *scaffolds* to aid the process of learning derives from the practice of constructing and repairing buildings or other large structures. The problem lies with the notion that teachers can ever know for sure just what kind of 'cognitive structures' are supposed to be built and that a stepwise series of instructions may be assembled that will result in a completed mental construction.

Why this alternative account matters

As science educators, I truly wonder about the degree to which our initial training in the natural sciences may have influenced the manner in which we perceive science education. It would not be inconceivable to consider how educators deeply immersed in the scientific worldview seek to develop science education in the shadow of our initial disciplinary instinct training. The question that has always been asked, and continues to be valid especially as we think about making, and possibly interdisciplinary STEM education, is the extent to which education is an art or a science. To be sure, this question creates a totally unnecessary false binary. It is perfectly possible to understand a process as being both/and, instead of a strict either/or. However, the concern for me is that under the suffocating neoliberal order blanketing large parts of the 'developed' world, common sense notions of what constitutes ideal organisational interaction have taken over. While there are many implications of neoliberalism in societies, for education, some of the more startling effects has been to reduce the educational interaction to one resembling business transactions. With standardised curricula and assessment methods to 'benchmark' schools against one another on narrow measures of outcomes, schools have been made 'free' and autonomous to direct their business in any way they see fit, as long as the required outcomes are attained (see, e.g., Smith, 2011). The scientistic/technocratic mindset can be particularly compatible with neoliberalism; in that a scientific reductionism and a general misuse of science lead one to conclusions which can support the neoliberal agenda. As for science education, a scientistic and neoliberal orientation might make a research agenda based on instrumental rationales more likely. While I have no studies that explicitly show the degree to which published science education research tends towards a more quantitative, more close-ended, reductionistic orientation, in my limited experience, the ambition for most science education research tends towards interpreting education in scientific terms.

With science education directed towards scientific goals, what could be lost is an attention to the humanistic goals of education. Education is not merely the effective communication of information or knowledge, even though these are very important. Although it can be very important to 'operationalise' scientific understanding, and measure growth in such understanding, education consists of more than transmitting and receiving knowledge as if between microwave antennae. Instead, education can be perceived, according to Biesta, as the creation of particular kinds of individuals, with particular orientations to phenomena and other human beings. The scientific ambition for explanation, prediction, and ultimately, control, may be acceptable and perhaps even desirable if we are dealing with inanimate objects. It becomes another matter entirely when we begin to cross the line into living things, and ultimately, other human beings. Even though we can, in most circumstances, claim that the acquisition of scientific knowledge in particular poses no threat to one's well-being, attempts to make education a science can tread awfully close to the distasteful end of the spectrum of possible human actions. Such attempts are objectionable on two grounds. Firstly, desiring control of human others violates the humanistic purpose of education: education should desire to expand the realm of possible human action, and not limit it; further, an education worthy of its name should take steps to accommodate the fundamental human quality of refusal. Secondly, especially for scientific knowledge that will have practical consequences in terms of human action, it is not clear that the scientistic, purely rationalist approach will be effective in generating desirable outcomes. This is especially in light of the phenomena of motivated reasoning (Kunda, 1990; Sharon & Baram-Tsabari, 2020; Sinatra, Kienhues, & Hofer, 2014), whereby it is understood that we are not cognising machines but actually make use of our emotions to derive preferred explanations for phenomena. Indeed, an education in science is as much an education of the aesthetics of what makes a desirable scientific explanation as it is the logical rules in which we use to distinguish accuracy. Consider how it is that we desire 'elegant' theories, or how theories equally supported by whatever limited evidence are chosen as canonical. Consider, for instance, how the geocentric model of the solar system is, for most purposes besides launching space probes, equivalent to the heliocentric model, but the latter is preferred only because it is 'less clumsy'.

Science education should not merely be considered as a transmission of content; there are values, preferred ways of interpreting phenomena within disciplinary lenses, and there are culturally specific ways of thinking about what ought to be done. What comes to mind are significant events of the last century such as the creation of the atomic bomb; but equally scary today are the gene editing tools such as CRISPR, or even the production of computer software—it is now completely accessible for middle- or high-school laboratories to perform gene editing and not to mention the ubiquity of computer intrusion tools. Whether or not students in such labs decide to simply insert bioluminescent genes into organisms, or merge infectious and virulent properties in certain viruses, is merely a function of how the lab culture 'works'. Perhaps not relevant to sites of formal instruction, where deviations from 'worksheet' style laboratory activities are rare, makerspaces need to be concerned with the kinds of pursuits considered desirable among the people occupying it. The argument here is that these sociocultural concerns need to be front and centre of people's minds as they plan for and implement these spaces. Makerspaces are not merely 'fun and games' to raise the level of engagement of students as if it were the sugar coating on a bitter pill of science and mathematics education; there is a serious purpose to organising students into collectives where there is some sort of education into the social organisations that produce better or worse outcomes, even if merely for conceptual gains. Here, the idea of makerspaces challenges the dominant ideology of learning as an intensely individualistic, 'meritocratic' process where one's success or failure in acquiring knowledge is purely a function of one's intellect. What makerspaces can be good for, if educators give it a chance, is the formation of knowledge collectives dedicated to investigating the relationship between phenomena and its associated representational claims that we conventionally communicate as so much propositional knowledge that, for some reason, *needs* to be learnt. Makerspaces are not a kind of 'educational technology' in the somewhat dated sense of a technology that helps people achieve the same goal in simply a more efficient way. Instead, I think of makerspaces as a challenge to the status quo of schooling, as a means of rethinking what it means to be educated, to understand science. Makerspaces should be a *cultural technology*, a different way of organising people for the task of learning, so that different goals are privileged.

The point of this chapter has been to highlight the importance of the kinds of new knowledge and new artefacts, and the sociocultural conditions that make it more or less likely that certain of these outcomes occur. Echoing the sentiment of the Black Lives Matter movement, it is not that there are 'bad apples' or 'mad scientists' who decide to make bombs, novel transgenic organisms, or social media applications; it is that there is a culture in place that makes it desirable to have these outcomes. Also contributing are funding models for scientists and technologists that make it especially likely that STEM practitioners are encouraged to ask no questions and approach their work as an amoral enterprise. With increasing specialisation, and by working in highly compartmentalised sections highly insulated from other groups, it becomes possible for researchers to lose sight of the 'big picture'. In this way, no party in particular ever becomes responsible for when things go wrong. In STEM education, it seems to me that we should not continue in this manner, obscuring the ethico-moral perspectives of STEM and pretending that science is an amoral enterprise whose outcomes are simply (ab) used by political agents who hold power over the hapless STEM practitioners. Students need to become familiar with the sociocultural aspects of their work, from the external relationships of STEM with society, to the internal sociocultural mechanisms that help ensure that practitioners police themselves. In this regard, the inclusion of technology and engineering into a transdisciplinary 'STEM' education at the school level brings with it several potential avenues for enriching science and mathematics instruction. As 'applications' of science and mathematics, technology and engineering are especially obvious exemplars of the ways in which STEM knowledge interfaces with societal needs. At the very least, makerspaces could be sites where students begin to understand how artefacts have been designed to meet socially desirable goals, and how these goals are but one possibility among many. Significantly, understanding that the means taken to achieve such goals as exemplified in a particular artefact or experimental set-up is merely contingent to peculiar conditions can be a powerful means to innovative thinking—if one understands the contingent basis for decisions, one is more likely to consider otherwise. For instance, the dominant model of the internet privileges an advertising model of funding, in which users apparently obtain free services in exchange for selling their attention to advertisers. Such a model by no means is ideal, as the potential for surveillance and manipulation has been realised over the last decade (Zuboff, 2019), often to disastrous consequences for both individuals and societies. Alternatives exist, say, in the model that is dominant in the Chinese internet, where users make micropayments in exchange for entertainment, but this can be accompanied with other problems as well.

In summary, this alternative account of how creation happens in science and technology is important because it begins to demystify the process of how innovation happens, such that STEM educators can have an alternative means to thinking about how to nurture creativity. Especially for STEM artefacts, experimental set-ups, and knowledge, growth is not a mechanical process in which some kind of ideal procedure can guarantee a good outcome. It is messy, unpredictable, and deals with peculiarities of the context, and can be the result of unique time and space conditions. This is not to say that STEM knowledge does not travel, but that we have to be careful when it does, and when it does not. Especially in our much vaunted 'internet age' where information is available within three clicks of a search engine, educators may want to rethink the importance of knowing facts, as opposed to knowing how to know facts. Given the ubiquity of information, what becomes the new scarce resource is how to put together new assemblages of information to form new knowledge claims, as well the *wisdom* to decide why such acts of creation ought to be. Again, such wisdom is not within the realm of propositional knowledge claims, to be implemented as strict rules that a purely logical cognising machine can interpret. Rather, wisdom comes from the collective thinking of a group of people with its specific emotional tenor, which can inspire such inventions as interplanetary space probes, invasive surveillance software for profit, or literal weapons of mass destruction.

How might such a makerspace work?

One such makerspace that focussed on the cultural aspect of its function occurred in Singapore, in Able Secondary School, as discussed in a previous chapter. Able Secondary School's makerspace entertained a group of about 20 students. These students were accommodated in a space that was formerly a workshop and its preparation room. The school's leadership appreciated that their students needed some form of technical expertise so that their students could design and build set-ups to conduct novel experiments that were not the textbook variety. This was important in order for their students to obtain practice in making new knowledge claims in response to the science that they were learning in the classroom. To reiterate, Able Secondary School had developed a programme to develop science and mathematics talent, and they quickly realised that a vital part of such talent was not the mere cognitive acquisition of conceptual understanding. Instead, especially for the natural sciences, being 'good with their hands' and understanding the relationship between phenomena and its representation was important. This resulted in the creation of a recreation club for students who found that they had a special passion for making things, which I will dub here the Innovation Club. To be sure, this naming is not official, and its function is closer to a homework club, but they certainly did attempt and create innovative artefacts, often not for any particular purpose than to satisfy their own desires. For instance, one student had learnt about the Tesla coil and wanted to experiment with one. At home, he had already built his own small scale tabletop model, but when he started to have designs on a larger and more hazardous floor standing model, his parents put a stop to that. Instead, he requested permission from Able Secondary to pursue such a project. Now, many schools would be risk averse enough to deny such a request, but Able Secondary's school culture of nurturing active experimentation in their students was somewhat tested in this regard. Tesla coils could reach dangerous voltages, and its discharge could give bystanders a nasty shock. To their credit, Able Secondary decided not only to support this project with material resources, space, and time, but also they decided to tap on their network of experts in the field to find someone who could advise the student on high-voltage circuits and electric discharge safety. The result of this was a Tesla coil about a metre and a half tall and was dangerous enough that it was not usually shown to students, at least not when powered.

The story here is not to do with this particular student, but what happened after. Although I do not have direct evidence for how this project inspired specific students, subsequent cohorts of students do talk in admiring tones about this student or the project. Even then, it was not quite so much a quiet reverence of a grand achievement, but rather a kind of bemused respect that he managed to get it done. Such an accomplishment gave the school a point-at-able exemplar of the lengths that they expected students to go, and also how much they were ready to support students' initiatives, however 'crazy' they might be. Along with this somewhat legendary project, students and teachers liked to tell me about other inspiring inventions and 'hacks', such as an unconfirmed episode where a student bought and modified a near field communication (NFC; contactless) card programming device to experiment with security protocols, a fan that did not have any moving mechanical parts, which was being considered for a patent, and many others. All these inventions and hacks were always spoken about in an excited, almost breathless manner, to express indirectly why the students and teachers were keen transgressors of norms and were willing to go the extra mile or do something out of the extraordinary. Even within my third visit to the makerspace, students had demonstrated how they were happy to push boundaries when, during a conversation, the topic of capacitor polarity was discussed. Electrolytic capacitors are designed such that their leads were to be connected to specific positive and negative voltages. A few of them had discovered the hard way the consequences of making a mistake with the connections; as I had not myself witnessed what these consequences were, they quickly wired up one small capacitor the wrong way to show me. Within seconds, with a smartphone video recording the event, the capacitor exploded with a pop that was disproportionate to its size. Everybody present at that time had a good laugh at the event, and the video was then circulated amongst students and their teachers. A few days later, just to show me how proud they were at how transgressive their students could be, their teacher sent me that same video, unaware that I was personally present at the time of the recording.

While I was present collecting data from the site, the students continued to surprise me with the level of creativity they would display, from Beverly's particular student project of electromagnetic propulsion (see previous chapter), to a student, who had a penchant of making improvised vehicles with modelling materials, a hobby flight controller, and a set of powerful motors. He would attempt to fly an autorotor model in the school field, which would result in the crash landing of the model at least a dozen times, the last fatal one breaking the body in half. Within a week, the autogyro model had been remade as a hovercraft and was now buzzing around the feet of the other students in the lab. Yet another student had collected a 'dangerous' (by his description) amount of inexpensive capacitors, connecting them in such a manner as to accumulate a lot of stored energy which he intended to release in a very short time. The only problem for his plans was that this collection stored so much energy that the transistor he wanted to use to control the pulse could not be purchased because of nuclear weapons' embargoes (which made him question if and why Singapore was on the embargo list).

Again, the point here is not to focus on the individual student achievements as examples of how Able Secondary managed to attract and nurture high performing students. What is significant here is the culture of the school and the makerspace in particular: with prominent examples, and recurring stories told of innovative science experiments that are retold (and therefore inadvertently exaggerated over the ages), new students are inducted into a community that is continually reminded that the essence of their learning science is so that they can one day surpass the performance of their seniors and mentors. This is decidedly different from, say, a community of scribes in medieval Europe whose main function was to ensure accuracy of reproduction, and in some very distant manner, this form of intellectual culture can be said to live on in conventional schools still invested in the banking model of schooling. Consider the event of the exploding capacitor: this event is not to be underestimated—firstly, it was taken for granted that the students could connect an electrolytic capacitor directly to a power supply unit, without worry that any of the components would get damaged. Even if, in the case of the capacitor, damage occurred, there would be no negative consequences for anybody involved. Secondly, students felt safe sharing video evidence of this event with their teachers, and teachers considered it a point of pride to share the video with a researcher from outside the school. This taken-for-grantedness is an indication of the cultural patterns of behaviour existing at Able Secondary and their makerspace community and is the result of deliberate decisions at every point where the policy is tested. Even trusting student leadership to self-manage the space poses a risk—of mess, of danger—that many other schools would rather not have to deal with. As a researcher, I have visited (too) many other schools whose laboratories and workshops have been emblazoned with very uninviting messages such as 'NO STUDENTS ALLOWED UNLESS TEACHERS ARE PRESENT' in capitalised letters and typefaces which connote that this command is strictly non-negotiable.

Instead, Able Secondary's makerspace was available for students to access at any time of the day. Because the preparation room had air conditioning, it was a welcome relief in tropical Singapore and attracted students who could 'hang out' in between their scheduled lessons to chat to each other about their day, and to catch some time to continue work on their projects. Besides the physical space, having a like-minded group of peers to reinforce a particular world view was useful not only for the group identity of the club, but it also helped students define for themselves what would be appropriate behaviour for one who was to acquire the identity of innovative student of science. For instance, Alfred, the teacher who was responsible for the makerspace, would come to the makerspace when his schedule could afford him time to do so, or when a student made an appointment with him to ask about a problem. When he did, students had a way of suddenly appearing to informally gather in a circle to chat. Such a chat need not be related directly to the projects the students were working on, although they very often did. He often lent a listening ear, asking the right questions that would quickly diagnose the most likely cause of the frustration the students were experiencing. However, what was more appreciated were not the overt 'teaching' talk or problem solving, but rather some 'idle chatter', where Alfred would tell them about something interesting that he read about that day, and students would take turns (in a completely unstructured manner) to talk about what interested them. This often resulted in flights of fancy, complaints about mathematics teachers who set impossibly hard to prove problems for homework, resulting in everybody pitching in ideas, and a generally open agenda for these sessions. Informally, Alfred modelled for the students the appropriate strategy for thinking about these invention problems, and especially, the appropriate goals that students should desire if their initial plans went awry. Interestingly, these sessions were not explicitly pedagogical in nature, in that there was no attempt to communicate a particular principle or theory, and all participants, students and teacher alike, could contribute ideas and critique each others' proposals. Rather, Alfred set the tone for how ideas should emerge and evolve; and what goals should be desirable. For instance, on one occasion students were experimenting with a device that made use of a surface tension phenomenon that some students had read about in an academic journal. They wondered how it could be put to good use in some practical manner, and it was during one of these brainstorming sessions that they decided that molten chocolate could be a good candidate, and how the experimental set-up was to be modified for some hypothesized properties of molten chocolate. When subsequently they tried the experiment for the first time, it was the 'expert eyes' of Alfred whom they relied on to understand why the experiment did not behave as expected, and what signs he relied on to make the judgment call of what to modify next. Here, the tacit knowledge acquired through years of having worked on practical investigations allowed Alfred the ability to have a 'feel' for phenomena and what was more or less likely to shift the experiment towards the desired outcomes.

As Alfred modelled this form of judgment to the entire group of students, the senior students were most likely to appreciate and understand what was going on. This way of 'learning to see' was then a skill the seniors could practice when the junior novices came up with problems that they could not troubleshoot. Indeed, when I asked Beverly (of the electromagnetic projectile weapon I introduced in an earlier chapter) about *how* she went about troubleshooting, she could not really explain it, beyond her understanding that it seemed 'obvious' to her, 'start with the most obvious cause, and then proceed to less and less obvious causes'. In this regard, this aspect of designing, tinkering, figuring out, and open-ended problem solving was the key learning that the students encountered in the makerspace community. As this 'feel' and 'eye' for phenomena was not something that could be easily reducible to a generalised concept, and was highly contextualised, it appeared that the best way to instruct students in this manner of creative problem solving was to simple involve them in increasingly sophisticated versions of it. Crucially, having more experienced students informally mentor the younger ones was a vital part of this process; organising a makerspace culture in which peer learning could occur was a key factor to the success of the makerspace at Able Secondary.

Summary

This makerspace at Able Secondary is certainly not an explicit 'STEM classroom' type set-up where students come in for programmed lessons. At the same time, this is not a 'hobby' type makerspace in a public setting in which the main concern is for a kind of recreation, and any learning was only to improve the enjoyment of the hobby. Examples of the latter include gardening or car restoration clubs, where participants do learn about their hobbies, sometimes acquiring very esoteric information and becoming experts at what they do, but do so purely as recreation. Able's makerspace enabled students to learn a very vital aspect of science—how to interpret phenomena and relate what was happening to the theoretical representations of high-value scientific and engineering knowledges. I argue that these processes have to occur in a relatively unstructured manner because of the nature of these kinds of learning: the interpretation of what counts as an ideal outcome of these investigations is not reducible to a simple principle or rule that can be applied to a variety of contexts. The key method of this form of learning has to be a kind of passive, background 'learning culture' that makes participants more susceptible to making the correct kind of conclusions from ambiguous phenomena which provides signs that need interpretation. This kind of culture can be created and sustained by firstly having a different orientation to the purpose of school: schools should not be merely a place for a form of reductionist transmission or even 'construction' of conceptual knowledge as if downloading a movie file across a microwave internet link. In fact, even if it were so that education is a transmission of sorts akin to digital transfer, it would still be important to ensure a commonality of transmission protocol and to ensure that the medium of transmission is optimal for the process. These factors find their analogue in humans as the sociocultural milieu that sets expectations of what is supposed to be happening in various contexts: for instance, that in silicon valley start-ups, 'disruptive innovation'; and in jihadi terrorist cells, disruptive innovation of a different sort.

Conclusion

In this chapter, I have problematised scientific knowledge—it is not a unidimensionally good form of knowledge that will automatically give us all manner of social benefits. Yes, every student deserves to know about science and technology, but it can also be very important to understand the ways in which science and technology may be used and abused in society. Far too often, science is presented in congratulatory tones as an essential knowledge for participation in contemporary society. While that may be true, a certain critical appreciation for how scientific knowledge is created, and the purposes to which we may deploy science and technology also need to be part of the science curriculum. The novel perspective here is the degree to which the production and deployment of scientific and technological knowledge is dependent on sociocultural processes, and how the processes of acquiring these skills are inherently messy and not easily (if at all) amenable to a transmissive pedagogy that disregards the fundamental humanity of the educative interaction. For me, such a humanity recognises the ability of people to refuse, to think otherwise, and to derive a future that can conceivably avoid the potential tragedies that await us if we are to adopt a 'business as usual' approach to the deployment of science and technology in society.

References

Amabile, T. M. (1983). The social psychology of creativity: A componential conceptualization. *Journal of Personality and Social Psychology, 45*(2), 357.

Amabile, T. M. (2012). *Componential theory of creativity.* Boston, MA: Harvard Business School. https://www.hbs.edu/faculty/Publication%20Files/12-096.pdf.

Biesta, G. (2004). Against learning: Reclaiming a language for education in an age of learning. *Nordisk Pedagogik (Nordic Studies in Education), 23*(1), 54–66. https://www.idunn.no/np/2004/01/against_learning_reclaiming_a_language_for_education_in_an_age_of_learning.

Biesta, G. (2013). Interrupting the politics of learning. *Power and Education, 5*(1), 4–15. https://doi.org/10.2304/power.2013.5.1.4.

Biesta, G. (2016). *The beautiful risk of education.* Abingdon: Routledge.

Brian Arthur, W. (2009). *The nature of technology.* London, UK: Allen Lane.

Caputo, J. D. (2006). *The weakness of God: A theology of the event.* Bloomington, IN: Indiana University Press.

Collins, H., & Evans, R. (2017). *Why democracies need science.* Hoboken, NJ: John Wiley & Sons.

Conti, R., Coon, H., & Amabile, T. M. (1996). Evidence to support the componential model of creativity: Secondary analyses of three studies. *Creativity Research Journal, 9*(4), 385–389. https://doi.org/10.1207/s15326934crj0904_9.

Dennett, D. (2009). Darwin's "strange inversion of reasoning." *Proceedings of the National Academy of Sciences, 106,* 10061–10065. https://doi.org/10.1073pnas.0904433106.

Dennett, D. (2013). *Intuition pumps and other tools for thinking.* New York, NY: Allen Lane.

Denzin, N. K., & Lincoln, Y. S. (2000). The discipline and practice of qualitative research. In N. K. Denzin & Y. S. Lincoln (Eds.), *Handbook of qualitative research* (2nd ed., pp. 1–28). Thousand Oaks, CA: Sage.

Denzin, N. K., & Lincoln, Y. S. (Eds.). (2018). *The SAGE handbook of qualitative research* (5th ed.). Thousand Oaks, CA: SAGE Publications.

Feyerabend, P. (1975). How to defend society against science. *Radical Philosophy, 11,* 3–8.

Firestein, S. (2012). *Ignorance: How it drives science.* Oxford: Oxford University Press.

Gambetta, D., & Hertog, S. (2009). Why are there so many engineers among Islamic radicals? *Archives Europeennes de Sociologie. European Journal of Sociology. Europaisches Archiv Fur Soziologie, 50*(2), 201–230. https://doi.org/10.1017/S0003975609990129.

Gambetta, D., & Hertog, S. (2016). *Engineers of Jihad: The curious connection between violent extremism and education.* Princeton, NJ: Princeton University Press.

Gunstone, R. (2013). [School] science, the learner, and the twenty-first century: What science? What learning? Fifth International Conference on Science and Mathematics Education, 34–49.

Harding, S. (1991). *Whose science? Whose knowledge?* Ithaca, NY: Cornell University Press.

Hayek, F. A. (1952). *The counter-revolution of science: Studies on the abuse of reason.* New York, NY: The Free Press.

Hutchins, E. (1995). *Cognition in the wild.* Cambridge, MA: MIT press.

Hutchins, E. (2014). The cultural ecosystem of human cognition. *Philosophical Psychology, 27*(1), 34–49. https://doi.org/10.1080/09515089.2013.830548.

Kunda, Z. (1990). The case for motivated reasoning. *Psychological Bulletin, 108*(3), 480–498. https://doi.org/10.1037/0033-2909.108.3.480.

Labaree, D. F. (2004). *The trouble with Ed schools.* New Haven, CT: Yale University Press.

Média., Le (2018, November 12). Avoid the apocalypse: Bernard Stiegler. Youtube. https://www.youtube.com/watch?v=3ggF2jE5d8M.

Liew, C. W., & Treagust, D. F. (1995). A predict-observe-explain teaching sequence for learning about students' understanding of heat and expansion of liquids. *Australian Science Teachers' Journal, 41*(1), 68–71. https://www.researchgate.net/profile/David_Treagust/publication/234752631_A_Predict-Observe-Explain_Teaching_Sequence_for_Learning_about_Students'_Understanding_of_Heat_and_Expansion_Liquids/links/56a9a07008ae2df8216539d2.pdf.

Luke, A. (2011). Generalizing across borders: Policy and the limits of educational science. *Educational Researcher, 40*(8), 367–377. https://doi.org/10.3102/0013189X11424314.

Nieusma, D., & Blue, E. (2012). Engineering and war. *International Journal of Engineering, Social Justice, and Peace, 1*(1), 50–62. https://doi.org/10.24908/ijesjp.v1i1.3519.

Patel, R., & Moore, J. W. (2017). *A history of the world in seven cheap things: A guide to capitalism, nature, and the future of the planet.* Berkeley, CA: University of California Press. https://www.ucpress.edu/ebook.php?isbn=9780520966376.

Reid, R. (2019, June 17). Ars on your lunch break: Let's talk about the extinction of humanity. Ars Technica. https://arstechnica.com/ars-podcast/2019/06/ars-on-your-lunch-break-lets-talk-about-the-extinction-of-humanity/.

Roth, W.-M. (2015). Becoming aware: Towards a post-constructivist theory of learning. *Learning: Research and Practice, 1*(1), 38–50. https://doi.org/10.1080/23735082.2015.994256.

Sagan, C. (1995). *The demon-haunted world: Science as a candle in the dark* (1st ed., p. xviii, 457 p). New York, NY: Random House.

Sharon, A. J., & Baram-Tsabari, A. (2020). Can science literacy help individuals identify misinformation in everyday life? *Science Education, 27,* 353. https://doi.org/10.1002/sce.21581.

Sinatra, G. M., Kienhues, D., & Hofer, B. K. (2014). Addressing challenges to public understanding of science: Epistemic cognition, motivated reasoning, and conceptual change. *Educational Psychologist, 49*(2), 123–138. https://doi.org/10.1080/00461520.2014.916216.

Smith, D. V. (2011). Neo-liberal individualism and a new essentialism: A comparison of two Australian curriculum documents. *Journal of Educational Administration and History, 43*(1), 25–41. https://doi.org/10.1080/00220620.2010.532864.

Soble, A. (1995). In defense of Bacon. *Philosophy of the Social Sciences, 25*(2), 192–215. https://doi.org/10.1177/004839319502500203.

Wieman, C. E. (2014). The similarities between research in education and research in the Hard Sciences. *Educational Researcher, 43*(1), 12–14. https://doi.org/10.3102/0013189X13520294.

Zuboff, S. (2019). *The age of surveillance capitalism: The fight for a human future at the new frontier of power.* New York, NY: Public Affairs.

6

THE INTERACTION OF HUMAN AND NON-HUMAN AGENCY

As I was preparing to write this chapter I found an interesting article that was published in the *Journal of Management Studies* which caught my eye. Seductively entitled 'A stupidity-based theory of organizations' (Alvesson & Spicer, 2012), the authors made a case that in organisational research, too much emphasis is placed on the role of intelligence, information, and knowledge, whereas what often needs as much attention is the role of stupidity. To be sure, these were not the first researchers to have paid attention to what has traditionally been ignored due to the negative connotation that ignorance and stupidity have on organisations; for instance James March (2006) pointed out that under certain circumstances, the foolishness of acting before thinking could be useful as a business strategy as it allowed for ideas which were not previously considered plausible to have a chance at being implemented. If we were to merely follow mechanical, rational decision-making, such ideas would never have been considered, let alone implemented. In any case, Alvesson and Spicer propose that the idea of stupidity in organisations is associated with three aspects: profound lack of reflexivity, justification, and reasoning only for myopic goals. If we consider the classroom to be the prototypical organisational unit that most people begin their enculturation into, to what degree to which are we welcomed into classrooms that teach us how to be reflexive, that remind us why we are doing what we are doing, and welcome substantive questions as to why all these questions matter (besides passing the gatekeeping examinations and getting on with our lives)?

But, in a way, much of the above is a digression. Stupidity is not necessarily always a negative property to be avoided at all costs; in some organisations, stupidity (as defined in the paper) can be beneficial to its efficient operation. Beyond the rhetorical question I pose above is a more serious question as to the nature of classroom culture that needs attention—if we can recognise that typical school classroom organisation prepares students for participation in stupid organisations,

DOI: 10.4324/9781351116220-6

how would that contrast with proud pronouncements everwhere that schools are preparing students for the knowledge industries? While these questions remain significant, this chapter wishes to discuss something else. Alvesson and Spicer assert that a primary reason why stupidity has dominated organisational culture is that businesses are increasingly focussing on symbolic manipulation, unmoored from the referents of these signs; in late industrial economies, much of what is being produced are goods which have no real immediate demand:

> This means organizations devote a significant proportion of their efforts to creating demand for their products by promoting expectations, producing images and influencing desires […] image intensive economic activity has become increasingly 'hegemonic' insofar as organizations engaged in extremely mundane activities are focusing significant proportions of their resources on image crafting activities.
>
> *(Arvidsson, 2006; Kornberger, 2010, Alvesson and Spicer, 2012)*

In science education, the technological revolution in the past half century has meant that an increasing amount of pedagogy has become virtualised—it has become possible, for instance, to model everything from atomic motion to population dynamics 'in silico', leading arguably to a more widespread understanding of the conceptual bases for the numerous theories that require evidence that is either very hard to gather or possibly too dangerous to implement in the typical classroom. Yet, we should consider historical precedent:

> [The] misuse of writing by the Sophists, according to Socrates, would consist in using it as a mnemonic device in a way that actually produces forgetting, where writing becomes a crutch that ultimately leads to forgetting how to think for oneself. The Sophists are those who sell this technique, the effect of which is to make the speaker feel as if they have all kinds of knowledge about many things when in fact they are losing the ability to know anything at all, and it is this feeling that makes this situation so dangerous.
>
> *(Ross, 2020, p. 466)*

The question I wish to pose here is the degree to which the focus on 'learning goals' in the purely conceptual sense can have a downside. In outline, I want to argue that humans are a tool using species, and while we tend to think of tools in a positive light of sparing us from back breaking effort, or in terms of affording us certain powers that the naked species has no access to, the effects of tool use are actually quite ambivalent. We now have planet level ecosphere altering powers; the dams we build destroy vast swathes of landscape and can actually change the period of rotation of the earth,[1] our effects are at a completely different scale to what beavers and other tool using species can have. If we can build mechanical systems to such an effect, we really should consider the ways in which the mechanisms of cognitive amplification and augmentation may be affecting the individual and collective mental landscapes. Unlike the previous chapter where I consider the goals that we might direct these tools to, in this chapter I want instead to think about the manner in which we (fail to) consider how tools are essential to our perception of the way the world is, how tools and humans mutually constitute phenomena, and crucially, how the material world 'pushes back' our attempts to make truth claims about it. The situation is as if non-humans possess a certain agency and behave in a manner that is non-deterministic; the Newtonian model of the well-behaved universe obeying eternal laws is likely flawed.

In other words, I am concerned here with what philosophers have termed as the ontological turn. While in an earlier chapter I have considered the implications of a misperceived and mis-presented epistemology for education and societies generally speaking, here I want to think similarly about the effects of a misperceived and mis-presented ontology. Ontology is the branch of philosophy concerned with what exists, the nature of such existence, and the nature of what is real, and what, in the first place, constitutes reality in itself. We often do not spend too much time thinking about these issues—things 'appear' before us in generally unproblematic ways; we often believe that the problems arise only in very marginal and esoteric circumstances such as in wave particle duality experiments. However, just as the concept of relativity advanced by Einstein upset the Newtonian model of the perfect universe, and was subsequently extended to a metaphorical understanding of the relative nature of truth claims (epistemological implications), a similar movement can be detected for quantum mechanics. Specifically, 'strange' behaviour at the quantum level has given us a new ontological metaphor and means of studying and thinking about what, exactly happens when scientists, technologists, or most generally, *people*, interact with things.

For science education, this chapter extends the earlier one discussing the social implications of a misperception of the epistemic processes behind how scientific truth claims are made. Recall, the tension is between an excessive faith in the apparently empirically grounded and therefore unquestionable evidence and an excess of distrust in the social nature of the making of truth claims. In epistemic terms, I would advocate a social realist position whereby we recognise the socially constructed nature of truth claims, but at the same time appreciate that such social construction is not completely arbitrary. With ontology, the question at stake for educators is the nature of the interaction between phenomena and its representation: if empirical investigations are not completely arbitrary, it must mean that, at one level, things provide equivocal evidence for some limited degree of latitude of interpretation. *How* do such equivocal behaviours arise, and what are the social rules that govern how we are to interpret such evidence? Also, just as with epistemology, widespread societal misinterpretation of the ontological nature of things, and the processes through which truth claims can be derived from our interventions with these things, threatens our collective future.

Consider an example. The recent disasters of the Boeing 737 Max passenger aircraft represent a terrible symptom of the problem that such a misunderstanding, a stretching of the limits of the ontological basis of science and technology past its breaking point. Investigators point to the manner in which the marketing and business divisions of Boeing led the development of the 737 Max as key to understanding why this aircraft suffered the problems that it did (Gallagher, 2019; Kitroeff, Gelles, & Nicas, 2019; Lefkow, 2019; Schaper, 2019; Stacey, 2020; Travis, 2019). As Boeing was facing competition from Airbus, the decision was made to push the development of the 737 Max, ignoring the concerns of the engineers involved. A decision was made to take an older, proven, airframe and to replace the engine with a larger, heavier, but more fuel efficient one. As many airports were already equipped with passenger handling facilities for older models of the 737 series, a decision was made not to adjust the height of the landing gear and render obsolete passenger handling facilities. As a result, the engine had to be placed on pylons which brought the engine ahead of the wings, so as to clear the ground. In the process, this new engine position brought about an instability in handling, which was apparently resolved by a piece of software called the Manoeuvering Characteristics Augmentation System (MCAS). This software relied on sensors, which were flight critical; however, for price considerations, this sensor did not have a backup. When sensors started giving erroneous information, as in the two crashes, the MCAS provided erroneous control override, resulting in disaster. What seemed to have made the situation worse is that Boeing had insisted that pilots who have had prior flight training in earlier models did not need any (costly) retraining and simulator time and could adapt to the new plane as if nothing had changed. When the erroneous MCAS system overrode the pilot, a lack of experience resulted in catastrophic loss of life.

Admittedly, this case is complicated and involves way too many 'articulating parts' to make a clean analysis. Nonetheless, it is still possible to identify the main problem is the manner in which humans have assumed that we can have control over nature. Intensifying since the advent of the industrial revolution, we have internalised a model of artefact construction whereby we assume we have enough theoretical knowledge to make arbitrary plans which will result in working artefacts. Hence, we see products often labelled 'designed in <country A>, manufactured in <country B>'; we think of design and manufacture as two distinct processes; and we think in terms of 'white collar' and 'blue collar' work. As we have no need of reminding, science and technology have amplified our capabilities that we are now a planet-level threat to the environment and, ultimately, to ourselves. Consider the possibility of minute errors in the match between our theoretical understanding of the universe (our representations) and the phenomena in question. Despite what we may teach or desire in classrooms, science and technology are merely 'good enough' approximations of reality. When scaled up, even minor mismatches will ultimately result in significant errors and unplanned consequences, as with climate change. The challenge for science education is to communicate clearly what we know, in relation to what we do not, and to give some clarity to the specific interactions at the boundary of knowledge.

This notion of control that humans have over nature, ultimately is one of the grand achievements of the European enlightenment, for which we appear to be beneficiaries of. We have committed acts of deicide: we killed the ancient spirits that inhabited our world, capricious things that foiled our ability to understand why things do not behave deterministically. But, as I suggest in the title of this chapter, we may have gone too far and either forgotten or learned to ignore the possibility that non-human actors can and do have agency too. This is not, of course, an attempt to bring ourselves back to some edenic past, but a careful appreciation and reconsideration of the nature of the human relationship with nature. I don't believe we actually communicate this relationship accurately in schools, and our reliance on educational technologies may actually exacerbate the problem. This chapter will study this problem and propose why makerspaces are a part of the solution, but only if we do educate appropriately. To begin, let me define the problem.

Hylomorphism

One entry point to understanding this problem comes from the close study of the '4A disciplines': Anthropology, Archaeology, Art, and Architecture. In Tim Ingold's (2013) analysis, these four disciplines share a commonality in that they collectively study or deeply involve the interaction of human culture and the natural world. For instance, anthropology and archaeology study how existing and extinct humans live. As a technological species (in the most general terms as generic tool use), we make arrangements of things and ascribe cultural significance to these arrangements which people outside of these cultures may attempt inferences of. Art and architecture are methods through which we may project onto matter certain configurations that are culturally significant. To reiterate, what we make, and what we find other societies to have made, have a particular cultural significance: artefacts do not randomly and spontaneously occur, they seemed to be planned according to the tastes and desires of the individuals and communities that the artefacts are made for. Because we are a social species, even individual design decisions are influenced by social norms of what an ideal artefact should be: either by attempting to exemplify and to become a better specimen of the norm, or in opposition to the norm.

One example may serve to demonstrate Ingold's approach to the problem. Consider the case of the Acheulean stone axe. This tool was one of the earliest remnants of our ancient ancestors and had been suggested as a significant piece of evidence for the intelligence of the human species. Found in three continents, and dated from over a million years ago, these axes appear to be deliberate pieces and share the same design. Unlike found objects with similar forms, these handaxes bear the imprint of a design emanating, one would imagine, from a culturally shared form. As Ingold reports, these handaxes bear the mark of the process of its making: formed out of flint, which has a tendency to fracture and break off in flakes when impacted at the edges where two surfaces meet. The making process leaves a characteristic irregular pattern of well-defined ridges [...] It is true that conchoidal fractures can occur accidentally, for example on the seashore when stones are knocked against one another in the surf. But Ingold asserts: "no accident, or series of accidents, could generate the systematic, patterned flaking of the handaxe." (p. 35).

Of interest here is the degree to which the design of these handaxes appears to have been unchanged over the million years or so that we know of its existence, and across the three continents where it has been found. To be sure, there have been slight evidence of 'progressive refinement towards greater balance and symmetry, through all that time the overall form remained virtually unchanged' (p. 35). It is hard not to think of the handaxe as a result of an intelligently conceived design being imposed onto matter in ways that reveal the intelligence of the species. Yet, as Ingold surmises, if the design of the handaxe is the result of a deliberate application of intelligence, how is it that there has not been any innovation over all that time? One candidate explanation has been that the handaxe has been the result of the body plan, an outcome of instinctual behaviour much like how beavers might create dams or termites make nests; each of these behaviours might result in complex structures but cannot be said to be the result of particular complex intelligence. Such a proposal puts archaeologists in a double bind, in that if the shape of the handaxe was a result of instinct or the unique manner the body interacts with matter, then we can account for the constancy of the form, but not the very obvious intelligence of the design. On the other hand, if the handaxe was the result of complex intelligence, then it is possible to account for the design, but not how the design has been unchanged over space and time:

> The roots of this dilemma run deep in the western philosophical tradition, through interminable arguments over the relation between the human body, understood as an integral part of the material world, and the soul that appears to bring to this world ideas and conceptions of its own. Ever since Aristotle, this distinction between body and soul has been taken as a specific instance of a more general division between matter and form. Any substantial thing, Aristotle had reasoned, is a compound of matter and form, which are brought together in the act of its creation. Herein [...] lies the foundation of the hylomorphic model of making. (p. 37)

Ingold also asserts that hylomorphism since the ancient Greek times has become more widely entrenched, and more unbalanced, with a privileging of the agentic mind, the creator of thoughts, plans, and designs, as the subject which can will ideas into being, and a deprecation of matter into a passive and inert object that simply receives these plans and has nothing to do but to be shaped into whatever the agentic subject desires.

Indeed, the solution to the conundrum of explaining the handaxe practically disappears if we stop thinking in hylomorphic terms. If we do not consider making the handaxe, or any artefact for that matter, as an imposition of a finished form onto passive and receptive matter, we can instead consider making artefacts as an emergent process. Such a solution becomes obvious when one carefully observes what modern day stone knappers do as they work the stone. Contemporary knappers will use a hard object, usually another stone that is harder than flint, often shaped like a rod, to strike at the edge of the flint. There is no known fixed quantity of force that one needs to strike at in order to achieve a flake, and there is no well-known relationship between the force, the angle of the strike, and many other factors, to the outcome of the strike. While it may be known *after the fact* that perhaps one struck too hard or at a wrong angle, very little predictive power going forward can be had of what is to happen with a processing step. This is very largely because the flint is not actually some passive material which has nothing to do but be changed by the processes imposed upon it, but it has within its structure, some form of stored tension and compressions arising from the manner that it had been formed by geologic action. The properties of the material need to be taken into account in the artefact making process; it is not possible to divorce the material property from the design of the artefact. While the human mind might intend for a particular outcome, it is not quite so much of an imposition of human will onto matter, but a respectful request: may I cause a change in this general direction, if I were to make this material handling step in this way? One never really knows until the step is taken.

As with stone, Ingold suggests that with some modification, similar considerations can be made for splitting wood with a contemporary axe made for the task of splitting wood. He cites Deleuze and Guattari (2004), whose analysis of such an act of splitting brings our attention to the manner in which the axe is guided by 'the variable undulations and torsions of the fibers'. Instead of the hylomorphic distinction between form and substance, Ingold asserts that the distinction, in processes of making, is between forces and materials. As I would add, even for a relatively pliant material such as clay, there is uncertainty in the manner in which it would react to human handling. Suppose I would be interested to make a clay artefact—I might have some notion of a plan for how the finished artefact is supposed to be, but despite whatever skill I may have at shaping clay, this batch I am working with may be too wet, or its adhesive property may be too weak, or it may react differently to the firing process than previous batches. I may have asked the clay to bend in ways that exceed its capabilities, and I will only know what this batch is capable of when I work with it. The only way to make, as Ingold cites Deleuze and Guattari, is to 'surrender' to the material and then 'follow where it leads'. As Ingold elegantly puts it:

> In this view, the process of making is a concatenation of separate steps that follow one another like beads on string. In the view we propose,

by contrast, the process of making is not so much an *assembly* as a *procession*, not a building up from discrete parts into a hierarchically organised totality but a carrying on—a passage along a path in which every step grows from the one before and into the one following, on an itinerary that always overshoots its destinations. Once again to adopt a helpful distinction from Deleuze and Guattari (2004: 410), this is not an *iteration* of steps but an *itineration*: making is a journey; the maker a journeyman. And the essential characteristic of his activity is not that it is concatenated but that it flows.

(Ingold, 2013, p. 45)

Ingold makes a well-considered case for all manner of interactions between human culture and ideas and the material world that we live in. As with archaeology, so it is with art, architecture, and anthropology: artists seldom create artwork with a fully formed idea of what it is they have in mind; while architects appear to plan a house, in its construction not all plans will be followed to the exacting details, and subsequent users of the house may not use the space as designed; studying how designers create designs blurs the notion of the separability of design and making.

Understanding how these disciplines approach the study of the interaction of human culture and nature gives us science educators a way to think about the relationship between the representations of science and natural phenomena. After all, science education is the process through which we inform students of our culturally agreed upon representations of how nature works. While we might believe that these representations are not arbitrary, but are based upon reproducible reductions of observable patterns of natural behaviour, we might want to rethink the limits to which we actually can have reproducible patterning of natural behaviour. Certainly, for some limited cases, we can have very good results, and for many other cases, a 'good enough' level of understanding is better than nothing, but the question for us is the level of confidence we ought to have on the general project of 'doing science', and what it is students ought to focus on when they are 'learning science'. To what degree do we actually have some form of deterministic, predictive, knowledge over how matter behaves—both in terms of the quantitative number of contexts and types of interactions as a fraction of the total range of possibilities, and also in terms of the qualitative degree of confidence we have in our knowledge? Reductive science places requirements on reality in order to make its predictions work: in physics, we consider point masses, frictionless surfaces, ideal gases, perfect circuits, and so on. In the messiness of reality, and especially in circumstances that matter such as in the angle of attack sensor of the Boeing 737 Max, certain boundary conditions may be exceeded in interactions which cannot be, even in principle, predicted. To somewhat belabour the point: also consider the nuclear disasters of Chernobyl and Fukushima, numerous industrial accidents such as Bhopal and Deepwater Horizon, or the numerous collapses of public infrastructure such as bridges and buildings. To what degree are we collectively culpable of a certain kind of hubris that we have somehow managed to 'tame' nature and placed harnesses on it much as one would tame a horse? When scientists take recourse in explanations such as 'chaotic behaviour' and 'non standard conditions', how much of these are the equivalent of us doing the equivalent of shrugging our shoulders and 'surrendering to the material and following where it leads'?

To reiterate: science relies on reduction, stripping 'extraneous' details from realistic contexts until a maximal, generalisable claim can be made about a minimised, idealised, and thus nonexistent scenarios. When we then make use of these principles to produce artefacts as in technological artefacts, we will attempt to shield the works from metaphorical 'sand in the gears', an act that may not always be possible. As science educators, to what extent should we communicate the ideal, as opposed to the real? As science educators, should we not also give students the rationales for why we chose to relegate particular aspects of reality as 'extraneous'? Should we not also give them a more accurate depiction of how science relates to the real world? In other words, should we stop communicating a hylomorphic model of science? In a manner that mirrors the challenge of communicating scientific epistemology to students, we are faced with a dilemma: if we teach students idealised versions of science, students might gain a powerful understanding at the risk of not understanding the limits of their knowledge. On the other hand, if we teach students realistic versions of science, students might come to appreciate how the universe actually behaves (in a not always deterministic fashion), but we risk misleading them about how effective scientific knowledge can be. Certainly, the situation is not so dire, and both the epistemological and ontological dilemmas can be somewhat resolved if we consider some form of age appropriacy of the messages that we might deliver. For instance, it might probably be less confusing to younger children to deliver the simpler epistemological and ontological messages, waiting until students are more mature before engaging in sophisticated thinking. However, doing so might disadvantage students who do not go onto more advanced training in science and lead to societies which do not effectively understand science. As Stuart Firestein (2012) suggests:

> There is a point in the training of every scientist in which they make the transition from textbook-oriented learning to discovery, not only of facts, but discovery of questions [...] Unfortunately this is precisely where the nonscientist gets left behind. Having been forced by an outmoded educational system to bear the misery of overweight textbooks, testing absent understanding, and valueless memorisation, they are left with this as their sole experience of science. Worse than the obvious distaste that this produces in so many, it also provides a disastrously distorted view of science as authoritarian, settled, and immutable. Fortunately, science is none of these things. But *who* besides scientists knows this?

Firestein's thesis is that there is a lot more that we do not know, than what we do. While school science, or school in general for that matter, is optimised for the telling of people various things about the world, we do a significantly worse job in communicating what we do not know. Certainly, trying to communicate an absence is hard and can plausibly only be achieved by reference to what exists. Among other things we do not know must be the non-deterministic behaviour of reality at the boundaries of our knowledge. Instead of a scientist position that presumes that all is in principle knowable and deterministically controllable, I propose that it is more important to communicate a position of humility, which recognises limits to our knowledge, and the careful means by which we can come to have knowledge.

Among other things we do not really know in science must be the emergent behaviour of collections and assemblages of things. To consider a classic example: just because we can understand the mechanics of a grain of sand, does not mean we understand how a pile of sand will behave. The pile does not behave merely as a linear sum of its parts, but performs very distinctly, requiring even a different kind of science to understand. As for piles, so it is with human contrivances, our devices and things: we test these artefacts and guarantee their function under perfect conditions, often just after manufacture. However, when entropy strikes and things break down after some time, we seem perfectly happy to simply discard them and to buy new ones that work predictably. Less common is the perspective that seeks to maintain, repair, or modify the function of the artefacts, to extend its useful life, or to essentially recognise the artefact as a thing that changes and behaves with a peculiar agency. Such agency, as with the flint, or the log, or even clay, may react at timescales different from what we are used to interacting with other humans. The material world does not repose in pure passivity in a state of suspended animation awaiting human intervention as if we are the only forms of agency that matter.

It is important here to note that I am talking about making and not manufacturing. For me, making is synonymous to creating, from whence we derive the positive qualities that we associate with creativity. I believe that making, as in creating for the first time, artefacts and ideas share in the quality that we can never know in advance what it is that we intend to create. There may be some degree to which we can intend an outcome, but it is seldom possible to precisely prespecify what it is that will result. Manufacturing is associated with standardisation and the elimination of variables to the state where as few surprises as possible can occur. But as we have seen from the possibility of industrial accidents, and the notion of 'manufacturing defects', nature has a unique manner of 'acting back' in response to human intention in ways that we cannot foresee. At the very least, I believe it is upon science educators to communicate how it is that we can have knowledge given the ways in which materials are not merely passive. Yes, science educators typically have students learn through practical investigations, but quite often, these investigations are designed to lead students towards the rhetoric of conclusions; less common are practical activities where students have to closely monitor what can be euphemistically dismissed as 'experimental errors' and to try to understand why experiments fail (Irwin, 2020). It is almost as if the only correct way to perform a laboratory investigation is to reproduce the standard result. Clearly, such a manner of investigation can only mislead students as to the nature of the human-nature interactions?

How is scientific knowledge related to experiment?

If Ingold has shown us how we need to acknowledge, if not actually 'surrender' to materials in our attempts to create artefacts, a possible criticism is that science may not necessarily be interested in making artefacts or that such forms of making are only a small aspect of science. To address this notion, I draw upon the work of Andrew Pickering (Pickering, 1995; Pickering & Guzik, 2008) and associates as they discuss the concept that has been referred to as the mangle of practice. In some way, this work may be seen as an intellectual offspring of sorts of Ian Hacking's (1983) earlier project calling into the question deep philosophical questions about the nature of theories and experiments. While I do not intend to summarise much of Hacking's arguments here, it is sufficient to note that his work marked a new focus on studies of the nature of science. The early positivists were interested in questions of demarcation ('why is scientific knowledge so special?') and focussed on how science was uniquely dependent on empirical confirmation. A second movement led by theorists such as Thomas Kuhn (1962/1996) started considering social factors that governed changes in scientific knowledge, or more generally, how scientific knowledge is influenced by the society of its practitioners. Hacking turned the magnifying glass on the practices of science; even the title of his book, *Representing and Intervening*, gives readers an indication of his approach. The question he began to ask was the reality of the theories and of the entities predicted by theories. Are electrons real, or are they a convenient fiction, and an artefact of the apparatus that has already been designed within the presuppositions of the theory to begin with, and thus likely to 'see' what it set out to? What role do experiments play in the adjudication of truth claims in science—are phenomena apparent for humans to observe, or must we coax nature to produce the sets of outcomes that we may infer the truth of theories?

For Hacking, an insight comes from the *failure* of experiments to 'work'. According to Hacking, undergraduate and high school students alike often report that courses in practical sciences are exercises in frustration as the mechanisms and theories reported about in textbooks almost never appear as promised in the laboratory. Even among research scientists there will be particular individuals who have greater or lesser facility in making even identical apparatus perform and produce the kinds of data that are necessary to empirically support particular truth claims. It is almost as if the old caricatured 'scientific method' whose first step was 'observation' is decidedly untrue, at least for the contemporary practice of science. Not only are observations theory laden, we don't actually passively 'observe' nature. Instead, we interfere with it, make it behave and act in ways that bring forth its particular aspects. That is the practice of science—the removal of 'extraneous' variables that might complicate matters and prevent our humanly generated hypotheses from being (disc)confirmed. What is pertinent, and what is extraneous, to an investigator doing any kind of science for the first time is not at all obvious and is not, even in principle, foreseeable. It constitutes something that comes close to a non-rational, perhaps even an artisanal approach, certainly not the hyper-logical, mechanistic reasoning method that we seem to portray scientific inquiry in student learning contexts. Hacking (1983) surmises:

> To experiment is to create, produce, refine and stabilize phenomena. If phenomena were plentiful in nature, summer blackberries there just for the picking, it would be remarkable if experiments didn't work. But phenomena are hard to produce in any stable way. That is why I spoke of creating and not merely discovering phenomena […] Noting and reporting readings of dials—Oxford philosophy's picture of experiment—is nothing. Another kind of observation is what counts: the uncanny ability to pick out what is odd, wrong, instructive or distorted in the antics of one's equipment. The experimenter is not the 'observer' of traditional philosophy of science, but rather the alert and observant person. Only when one has got the equipment running right is one in a position to make and record observations […] The pre-apprentice in the school laboratory is mostly acquiring or failing to acquire the ability to know when the experiment is working. All the thinking has been done, all the designing, all the implementation, but something is still missing. The ability to know when the experiment is working includes, of course, having sufficient sense of how this artifice works in order to know how to put it right. A laboratory course in which all the experiments worked would be fine technology but would teach nothing at all about experimentation (p. 230).

Consider, for instance how Robert Milikan was reputed to have thrown out pieces of data as spurious when he was conducting experiments to discern the quantisation of electronic charge. Surely there had to be extra logical reasons for his doing so, or he would not have faced accusations of scientific fraud. Eventually, his results were confirmed by other experimenters, but this only confirms the point here: there is something beyond the logical and rational that is at work when scientists perform their experiments. Perhaps, as Hacking suggests, experimenters need to be particularly attentive to their apparatus, finely attuned to the circumstances in which an experiment can 'go wrong'. Such an ability, a skill, need not be made mystical, to invoke such terms as individuals who are capable of mechanical sympathy, or even at an extreme, some sort of talent in making the machines 'sing'. But it certainly exists as a particular talent that cannot be easily reducible to a set of propositions that can be expressed in some form of representations. In other words, it is unlikely that one can communicate, *en-masse*, such forms of skill and understanding. As educators, we return to the classic conundrum of the nature of knowledge, and the process of communicating it: is knowledge something that is neatly package-able, into media such as books, multimedia productions, or even contemporary forms such as interactive digital games? Or is knowledge something more tacit? With echoes to Gilbert Ryle (1946) and Michael Polanyi (1966/2009), and of course, my earlier chapter, it is clear that there are at least propositional knowledge claims (knowledge that), and tacit knowledges (knowledge how), each of which have its own merits. If knowledge is easily delivered as in a lecture, education would be relatively inexpensive and efficient. On the other hand, if knowledge were to be acquired only through the lengthy process of apprenticeship, much of public education may actually be futile. This is especially in the earlier grades where students seldom get opportunities to experience the fuzzy boundaries between phenomena and its representation.

To be fair, and without forcing a false binary here, both approaches are needed and useful. But to repeat Firestein (2012) the concern here is for the students who will not go onto undergraduate or higher levels of instruction where students are typically provided with greater opportunities to be apprenticed into the expert 'ways of seeing' (Goodwin, 1994), ways of feeling, and ways of *being*. Without such exposure, it is more than likely that the general level of scientific literacy in society will be such that few actually understand how exactly scientific ignorance arises, or in the first place that the regions where we have knowledge are actually built upon rather tenuous ground. We inculcate in young children the possibility that our knowledge can be absolute and not subject to the whims and fancies of an unruly nature that our ancestors believed to be the act of spirits. We raise them to expect certainty, and the safety that apparently derives from the illusion that we have knowledge, we can predict all future outcomes, and we cannot be surprised. Yet, when the 'rounding errors' do not cancel each other out, when 'experimental errors' are actually unknown 'physical effects', when the unexpected happens and when crises occur in our understanding of science, it is often scientists that celebrate and find opportunities for learning. However, for those who have not been introduced to the accurate scientific way of thinking, they will be likely be led back to other prophets who promise safety and security, just the way that they were taught as children. In some cases, actual religious prophets intervene in these lives and provide a rationale and an all too alluring narrative as to who is to blame for the mismatch between the state of affairs on earth and in utopia, to disastrous consequences for societies (Gambetta & Hertog, 2016). Even without going into the rather extreme cases such as this, and without sufficient evidence at this point in time to make a stronger case, I really have to wonder the degree to which such misunderstanding of the ontology of science contributes to such crises as climate change, vaccine deniers, and socially irresponsible attitudes towards the pandemic.

Dance of agency, posthumanist performativity

Evidently, there is a problem with what I might call the gap between the phenomena and its representation. Phenomena are, to a large degree, *made*, or at least, captured, and just like the difference between wild animals in nature and animals in captivity, how captive animals behave can serve as a good approximation to how wild animals do. I expect that most readers might find this analogy just past the level of acceptability; after all, we have become used to thinking of matter as 'inanimate' and therefore without agency. But, even as we have seen with Ingold's analysis, in making artefacts (and by extension, phenomena) we have to contend with the manner by which matter retains stored compressions and tensions and disrupt the intentions of those who attempt to make. In any case, as Hacking opened the door for investigators to more closely examine scientific practices, others have stepped in and produced more substantial work. In this section, I will survey the work of Andrew Pickering and associates, who study the practice of scientific research from an anthropological lens. Pickering begins his analysis by upending prior assumptions about the study of the development of scientific knowledge. Prior to Pickering's analyses, philosophers and other social scientists of science have been content to study the end products of scientific practice, such as the knowledge claims, the theoretical products of science. Such an approach uses scientists' reports of what they said they did, as opposed to what they actually did. Already deleted from these reports are aspects of the scientific practice not deemed pertinent to the making of these claims. Such an approach is ironically, not scientific: if we are interested in characterising and discerning patterns of human social behaviour in science, relying on others' reports should surely not be sufficient. Pickering (1995) argues that we need to consider wider sources of evidence:

> Within an expanded conception of scientific culture, however—one that goes beyond science-as-knowledge, to include the material, social and temporal dimensions of science—it becomes possible to imagine that science is not just about representation [...] The point is this: Within the representational idiom, people and things tend to appear as shadows of themselves. Scientists figure as disembodied intellects making knowledge in a field of facts and observations (and language, as the reflexivists remind us). But there is quite another way of thinking about science. One can start from the idea that the world is filled not, in the first instance, with facts and observations, but with *agency*. The world, I want to say, is continually *doing things*, things that bear upon us not as observation statements upon disembodied intellects but as forces upon material beings. (p. 6, emphasis in original)

Pickering offers us what he terms as the performative idiom of science, studies that emphasise science as it is being made, with attention to the people and things and their mutual interactions. To be sure, Pickering's approach is not unique; Actor Network Theory (ANT) (Latour, 2005) had started to consider the role of objects as elements of social networks alongside human agents. Pickering suggests that treating objects, specifically scientific instruments, or machines in general, in this manner is appropriate as most of contemporary science no longer resembles the 'classical naked-eye astronomy' of old. Instead, scientific instruments have co-evolved with human scientific knowledge to amplify our senses and to improve both the accuracy and precision of our measurements. Inducting one into the fraternity of science has never been merely the acquisition of knowledge claims of the practitioners, but especially in contemporary times, an experiential initiation in the particulars of using scientific machines to make knowledge claims. Such a notion should not be surprising; for many students, an early laboratory experience is the use of basic measurement instruments such as vernier calipers and measuring cylinders or microscope reticles. While ANT approaches might consider such instruments on an equal standing (on analysis terms) with the human actors, Pickering's notion of the mangle of practice would not. Instead, Pickering asserts that there are important practical differences between material and human agency such that the actions of machines cannot and should not be directly comparable to humans. Humans and machines are 'constitutively intertwined; they are interactively stabilized' (Pickering, 1995, p. 17): machines require a particular set of operations in order to function properly, and a disciplined sense of human agency is necessary. Correspondingly, it is obvious that humans cannot serve as substitutes for the machines. Nonetheless, it still remains that humans can display and assert intentions, in ways that are not accessible to machines. While machines and materials cannot be said to have intentions, this does not mean that they are passive. As with the criticism of hylomorphism in the previous section, material agency is an emergent property that describes that essential unpredictability of the human-material interaction. According to Pickering, material agency:

> is *temporally emergent* in practice. The contours of material agency are never decisively known in advance, scientists continually have to explore them in their work, problems arise and have to be solved in the development of, say, new machines. And such solutions—if they are found at all—take the form, at minimum, of a kind of delicate material positioning or tuning, where I use "tuning" in the sense of tuning a radio set or car engine, with the caveat that the character of the "signal" is not known in advance in scientific research. (p. 14, emphasis in original)

Again, as with Ingold's critique of our conventional interpretations of making (artefacts), Pickering's notion of the mangle of practice similarly presents us a novel perspective to thinking about what it is that is being done when humans interact with materials to make, in the case of science (and technology), novel truth claims about the world. In both making artefacts and truth claims, there is much to be said about how human interactions with material are not at all a simple, predictable, deterministic, or otherwise controllable process. In what follows, I will review two examples from the work of Pickering and associates which dramatically demonstrate features of what Pickering termed as the mangle of practice. Among other things, what I have chosen to focus on are aspects of the mangle which I believe are especially important to science educators. These aspects are: (i) the essential unpredictability of material interactions as part of scientific practice; (ii) the quality of the work that is necessary to 'tune' instruments such that phenomena emerges; and (iii) the concept of resistance and accommodation as moves of the metaphorical 'dance of agency' between human and non-humans.

Obvious examples from living things and complex systems

It is perhaps not surprising to find the most obvious examples of non-human agency in live organisms and in complex systems. Especially for the latter, the scientific consensus has been that even though the governing principles of complex systems may be completely known and simply described, the iterated interactions between multiple elements of the system will result in unpredictability past a very short timespan. While we have been getting better at computational modelling of such systems, small initial perturbations can lead to large deviations in final conditions. How well we are able to measure all of the possible initial conditions, and how well we are able to discern the factors that matter, and which can be safely ignored, can in certain circumstances become literally matters of life and death. In Pickering's (2008) telling, the Mississippi River is one such complex system in which the human ambition for control appears to have been repeatedly thwarted by the non-human agency of the river flow. The US Army Corps of Engineers have been tasked with the responsibility for controlling the Mississippi, a task that it has taken on as a 'battle with the river—a battle in which the levees are central and whose outcome is far from certain' (p. 6). As the Mississippi is about ten metres above a neighbouring river into which it feeds (the Atchafalaya), it has a tendency to shift its flow. Doing so would result in the Mississippi reaching the Gulf coast over 200 kilometres west of its current exit, leaving large cities like New Orleans out of reach of its waters that the city needs for survival. In the service of this project of controlling the Mississippi, a 250,000-ton weir was constructed in 1963. The aim of this project was to keep the loss rate of the Mississippi to the Atchafalaya to its historic rate of 30%. Unfortunately, flooding in 1972 and 1973 almost destroyed the control structure—parts of the structure suffered massive damage, including holes as big as football stadiums excavated by turbulent flows. Repairs were made at 3.5 times the initial cost of construction, adding 2,600 tons of new gates and materials. Despite these modifications, an engineer working on the new project was reported as saying 'I hope it works'—such was the degree of confidence in their intervention.

The human intention in this case is to control the river, in a 'human-centered, atemporal, detached, control project'. Indeed, the Army Corps of Engineers actually intends to 'stop time', freezing the flow of the river to a particular point in time. However, as Pickering surmises:

> This project has always itself been embedded in a decentred and open-ended becoming of the human and the nonhuman, a "dance of agency", as I would call it, between the engineers and the river. The human agents, the engineers, try something—raising the levees, say—and then the non-human agent takes its turn by rising still higher and flooding New Orleans. In response the humans do something else—building the weir between Mississippi and the Atchafalaya—to which the river does something else—ripping and tearing away at it. And so on, forever. (p. 7)

Pickering asserts that this kind of behaviour, the essential unpredictableness of the 'real world', is not a property of rivers or (as I will elaborate below) living beings only, but a generalisable condition of the world. What is of interest here is also the fact that the human agents in this example are not acting blindly or by faith. The Army Corps of Engineers are a well-funded organisation which has been informed by the best science available in order to find 'the timeless hidden essence of the river and hence dominate it, first conceptually and then materially'. This science includes a six hectare scale model of the Mississippi basin, filled initially with walnut shells as replica riverbeds, until rot set in and then they were replaced with lumps of coal. For Pickering, this case exemplifies the dual nature of scientific knowledge: while it changes with time in response to the emergent behaviour of the natural world, it portrays knowledge about nature as timeless, and nature as eternally unchanging. In this manner, what science *is*, rather than the epistemic accuracy of its truth claims, is the focus of Pickering's inquiry. Rather than science as a tool that, as conventionally portrayed, illuminates and empowers, we need to acknowledge this ontological metaphor of science (as timeless truth) obfuscates and renders societies, if not individuals, particularly prone to disillusionment (and subsequent deception by unethical parties) when scientific knowledge inevitably updates itself. As with Pickering's use of Heidegger's observation that science is 'at best in the domain of the "correct" rather than the "true"'—could we science educators do a better job at communicating a more accurate ontological metaphor of science? While Pickering does not make any specific recommendations for educators, he does suggest a political and philosophical project that provides much work that needs to be done by all, including educators. Contemporary societies, run on Western ideals such as 'dualist detachment and domination' (p. 13), have since intensified this illusion because of Industrial Revolution models of material handling and production. The success of progressive industrial revolutions (steam, mass production, electrification, electronics, automated production, information, etc.) has more deeply entrenched this method of thinking about the passivity of materials and our apparent dominion over nature. Our material existence is comfortable, productive life expectancies are high, and it can become easy to believe that science and technology can give us a guarantee of a future of deterministic predictability.

Pickering suggests instead that we could do better with a 'politics of experiment', currently occupying the margins of our culture, in such organisations as 'the New Age movement, non-Western spiritualities, and cybernetics'. If the spell of determinism 'could be broken, the world in general would then strike us in our everyday lives as what it is—a place of decentred human and nonhuman becoming—and we would surely live very differently were that to be the case, self-consciously in the flow of becoming rather than denying it' (p. 13). As educators, we can become gripped within the illusion of dualist detachment and domination. It can be easy to think that what is important is to achieve a form of mechanistic efficiency of learning, of especially well-established theoretical propositions. We can forget that a certain kind of wandering can be educative, and we may even fail to remember that even the process of education can be, and in fact should be, a process of emergence. Instead, education research is more often than not preoccupied with studies purporting to demonstrate the effectiveness of some method or another to optimise 'learning'.

To return to examples of the mangle and attempts at understanding scientific knowledge and practice in terms of becoming rather than as a static body of timeless knowledge, another fairly obvious case derives from studies of living organisms. In this case, however, because these organisms are trees, the obviousness of arboreal agency may not be apparent (Franklin, 2008). This case concerns the Australian gum tree, the eucalyptus, which is the quintessential Australian tree. The eucalyptus is a unique plant in that it is fire resistant and in fact has co-evolved since the arrivals of humans on the island continent. Through accidental or selective deployment of fire, eucalyptus plants have come to cover the continent by surviving through fire events that kill off other competing species. Not only are eucalyptus trees fire resistant, they actually need fire for reproduction—their seed cases are durable and require a fire event to burn through a waxy barrier, before the seeds are then dispersed onto newly cleared land that gives the seedlings a competitive advantage of an absence of canopy. As an adaptation to fire, eucalyptus trees also regularly shed their bark and leaves, all of which containing a volatile oil; these trees also do a good job of drying out the soil and the underbrush, all of these seemingly 'designed' to catch fire at the slightest provocation. The first humans to have lived alongside these trees are the aborigines, who, along with their occasional tending of the forests by setting 'cooler' fires that cleared the accumulated shedding of the trees, also practiced a nomadic lifestyle ready to react to changing living conditions. When the colonialist European settlers arrived, they brought with them sensibilities and ideals about permanent settlement, pastoral living, and a suburban lifestyle of 'living next door to nature'; as well as a 'zero-tolerance' policy of forest fires. Inevitably, when one lives right next to eucalyptus forests and insists on a static impermanence that encourages the build-up of flammable forests, it becomes only a matter of time before a fire breaks out that becomes unstoppable, bringing with it overwhelming property damage, if not also a concomitant loss of life.

As Franklin surmises from this example, we tend to see trees as non-agentic, by their rootedness to the ground and their slow movement at best akin to the microscopic movements of the hour hand of a clock. Yet, a closer examination of the situation should make it 'clear that the social times of specific forest policies, management plans, fire events, and resource use are often out of kilter with the longer-term ecological and glacial times over which the trees and forests are active and influential' (p. 43). In having co-evolved with the aboriginal practices of forest management which features a more involved process, and then having humans abruptly (in evolutionary timescales) changing to a 'human hands off' practice, we may have disrupted the mutual dance of agency in ways that are (only) in hindsight predictable. Non-human agency exists, not as in humans that we understand as having intentions that can be expressed in some form of language or other form of representation. Neither is there an attempt to impute some sort of spirituality animating the non-human realm. Instead, non-human agency should be understood merely as a different interpretation of the notion of agency: just as we humans are capable of non-deterministic actions that do not flow from some scientific law, the behaviour of plants, rivers, and nature in general is not in practice, if not in principle, predictable in advance. That we can have scientific knowledge of the world, that such knowledge appears to give us certainty, is largely illusory, in that such knowledge depends on an 'interactive stabilisation' between human and non-human agents. Under certain known and well-established circumstances, with 'confounding variables' removed, scientists can make truth claims, temporally and contextually limited.

This is not to say that these knowledge claims are false, untrue, or incorrect, but that they are reliant on very specific configurations of the material world. That airplanes fly, or computers process information, and vaccines prevent disease are the outcomes not only of our scientific knowledge, but also the application of these knowledges within extensive degrees of redundancy and security. For instance, commercial aerospace equipment are built to safety factors that mean that they can easily bear double the typical load expected in normal operations. However, as dramatic footage of test rocket explosions can attest, or as with the Boeing MCAS failure, the moment we push the envelope, all bets are off. Again, as with Pickering, our notion of science and scientific practices tends to obfuscate the dimension of time from our understanding of how the natural world is. I would add that the conventional scientific worldview encourages a human detachment and discrimination from nature as if it were a passive recipient of human agency and cultural practices. Yet, as we consider from this case, we humans do not stand apart from nature at all. Especially for the living biosphere for which we are reliant for life support, a better way might be to think of ourselves as a unified natureculture, mutually interdependent and locked into a common destiny.

Not so obvious examples from the physics of subatomic particles

If the examples above can clearly illustrate what Pickering refers to as the dance of agency, it may also be the case that these examples involve somewhat animate objects and organisms for which some form of agency can be imputed, if only not at the same scale as that we are used to experiencing with other human agents. It may be acknowledged that such an interpretation merely stretches the definition of agency, but then, that may precisely be the point: we need to expand our interpretations and understanding of what acting with agency constitutes, and the diverse ways in which agency may be exhibited, especially at timescales different from what we may be used to. Pickering's idea of the mangle of practice is also constituted by the concept of *tuning*, the iterative steps one would make in the conduct of experiments in order to respond to the ways in which humans and the natural world come together to create a set of truth claims of value.

His earlier examples that introduced the notion of the mangle involved the close examination of the performative practices of particle physicists as they worked on detecting the subatomic particles and asserting certain theoretical propositions. His examples have to do with the work of Donald Glaser, as the inventor of the bubble chamber that allowed the visualisation of the traces of the subatomic particles as they passed through it. This work is significant, as it won Glaser the 1960 Nobel Prize, and the use of the chamber also contributed to the winning of another. Generally, as with the image of particle physics that we are accustomed to today, the early work that Glaser and others were doing consisted of beams of particles travelling at high speeds, impacting onto a target, producing a subsequent shower of particles. The effect has ever been likened to causing two cars to collide at high speed and performing a form of forensic study of the wreckage to establish the technical specifics of the cars involved. Unlike the metaphorical equivalent however, subatomic particles produced in these collisions may not leave a persistent wreckage. Instead, these particles may be short-lived and decay without a trace. In order to find out the particle characteristics such as its mass and electric charge, experimenters have these particle fragments pass through a bubble chamber, where the particles will interact with the vapour and leave visible traces that can be subsequently photographed for analysis. Glaser sought to improve the frequency of particle-vapour interactions by using a denser material for the vapour medium, but initial attempts to find a suitable material all failed, one after another. Subsequently, once a suitable bubble chamber was created, Glaser's work was initially with cosmic rays, whose arrival at the chamber was random and unpredictable. To get around this problem, an electronic triggering device was constructed that reacted to potentially significant cosmic ray events to capture photographs. Once more, this strategy resulted in several failures, for which Glaser devised a series of responses before he ultimately abandoned that strategy of seeking particle events in cosmic rays.

Instead, he shifted his focus to particles produced by accelerators, but eventually moved away from this study as he wanted to avoid the 'big science' teams of scientists and engineers working in interdependency and consequently somewhat reduced autonomy. What is interesting here is also the mangling of psycho-social factors behind the decision-making of science: because Glaser's competitor at that time decided to use liquid hydrogen instead of ether as the bubble chamber media and because hydrogen has a smaller mass than ether, a much larger volume of hydrogen needed to be used. As a result, it became necessary to employ large teams to work on the equipment and to coordinate their efforts, a particular line of action which was anathema to Glaser, and which therefore changed the direction of his research.

While it can be easy to overlook these events in scientific practice as merely the typical kinks in the practice of science, for Pickering, these aspects of Glaser's work demonstrate important concepts from which we may obtain meaningful advances in our understanding of the ontology of scientific practice. Prime among these insights is his notion of the mangle, a description of scientific practice in which we can understand how the material, conceptual, and social interact in emergent and unpredictable ways. As with the examples prior to this, the case of Glaser attempting to create a bubble chamber provides very clear demonstrations of the dialectic of resistance and accommodation, and the mutual dance of agency undertaken by experimenters and the material universe. Glaser sought to develop one particular configuration, based on the best knowledge that he could muster up at the time. Yet, the experiment failed; the natural world refused to cooperate. In Pickering's terms, this constitutes a resistance to Glaser's intentions, for which he now has to act in ways to accommodate this resistance. As Pickering emphasises, such material, non-human agency is not the sort that exists independent of human intention. The natural world is not filled with capricious spirits acting in non-deterministic ways. However, it is possible to have an *interactively stabilised* relationship with materials, where, if we continue to metaphorically dance together, it may be possible to sustain some degree of predictability as suggested by the representations that we call 'science'. In situations where such stabilisation has not yet been achieved, or in situations where stability is left to decay, it should not be surprising that the regularities we can become accustomed to break down, with undesirable consequences in many cases.

Pickering clarifies this notion of non-human, material agency as follows:

> material agency is bound up with that of material agency. Material agency *does not force itself upon scientists*. There is, to put it another way, no such thing as a perfect tuning of machines dictated by material agency as a thing-in-itself; scientists, to put it yet another way, never grasp the pure essence of material agency. Instead, material agency emerges via an inherently *impure* dynamics that couples the material and human realms. (p. 53, emphases added)

In other words, it can be suggested that material agency is closer to a metaphorical description of the process of working with material and not a literal description of a particular kind of 'intelligent' agent that can intend an outcome. There is no attempt here to return us to a pre-Enlightenment state of faeries and sprites that haunt rivers and the wooded lands, and malevolent gods that rain punishment from the heavens. Nonetheless, the key insight here is the impossibility of taming 'nature' and the form of agency here is the kind that needs to be imputed when we interact with other humans and things in ways that can defy our expectations of how they are supposed to behave. Just as human agency is apparent when we can make decisions virtually independent of the conditions that we are exposed to, material agency is implied when things do not always behave in a non-deterministic manner in response to conditions. Significantly, simply because we cannot impute some sort of intelligence or will to material interactions cannot be a means to dismiss the possibility of material agency.

Pickering's analysis also gives rise to the concept of a temporal emergence of intentionality—scientists do not go into the field with the full intention to prosecute a particular course of action, just as materials do not act with intention to frustrate scientists' attempts at arriving at interactive stability. While it cannot be fully said that scientists do not really know they are doing as they are doing it, it is also hard to deny that they are 'making it up as they go along': in response to inexplicable behaviour from experiments, scientists can change their apparatus and their underlying theories explaining what is going on. Human goals in scientific investigations can and do change in response to temporally emergent material resistance to human agency. What this means is that human goals and human intentions are not the sole determinant of the intention and direction of scientific investigations.

But what might this buy us?

To summarise, Pickering's mangle of practice introduced such concepts as human and material agency, non-human agents presenting resistance to human intentions, and human agents making accommodations in response. Pickering's conception is not the only one; since his work, or more generally, Ian Hacking's focus on the practices of science over its representations, the field of research on science-technology-society has developed what might collectively be known as a posthumanist performative orientation or ontological metaphor for science. Much of these developments have also emerged as part of a feminist response to the excesses of what was perceived to be a masculine domination emanating from science (see, e.g., Barad, 2003; Harding, 1991; Longino, 1990). Typically, arbitrary decisions would be made for the convenience of the typically male majority in decision-making; but the arbitrary nature of these decisions are obscured on the grounds of 'impartial' technical decisions. While I do not intend to delve too deeply into the problems raised by feminist scholars, I do acknowledge the significance of the philosophical advance represented by a shift in attention to the practices of science. It should be apparent that much of the problems that we face today stem from a kind of hubris that arises from a belief that we actually have some form of arbitrary power over the natural world. While it is probably not clear just what fraction of people accept such an interpretation of the authoritativeness of scientific knowledge, it is fairly common to accept that the solution to many of our problems will come from scientific and technological advances. The contention of researchers who oppose a technological determinist position may agree that what has contributed to this line of thinking has been pronouncements by early philosophers of science such as Francis Bacon:

> For you have but to hound nature in her wanderings, and you will be able when you like to lead and drive her afterwards to the same place again. Neither ought a man to make scruple of entering and penetrating into those holes and corners when the inquisition of truth is his whole object.

While this text comes from an article that defends Bacon against feminist charges of the masculine domination of science and its effects on society at large, I am choosing here not to take sides in this debate. Rather, it is my intention here to simply note that current problems that we have seem to be better understood as driven by such a Baconian model of science and the knowledge that we may generate. Take, for instance, climate change and the unsettling of the delicate ecological balances on earth—the spread of industrialised models of growth, based on the success of small-scale experiments that can afford to ignore 'rounding errors' and other departures from ideality, has ultimately resulted in the mass accumulation of such errors that our collective livelihood is being threatened.

Of course, we may gladly acknowledge that we now know better, that the science has changed, but it is most unlikely that we will proceed from these experiences with an expanded sense of humility that even our best knowledge is limited in ways that we cannot yet foresee. It seems what we need are new ontological metaphors for science: what *is* science and what are the legitimate forms of inquiry that scientists perform? If the Baconian metaphor for science is one of domination and control once the vital essence of nature is discovered and harnessed, the posthumanist metaphor for science instead presents a more equal relationship, dancing with nature instead of hounding 'her'. The posthumanist account, in my opinion, is not only more accurate, but it also has the potential to liberate us from our more destructive relationships with nature. In thinking about science as a mutual relationship with an agentic nature that requires our continual partnership, we gain the valuable perspective of decentring the human role in nature. Just as we have shifted from a geocentric model of the solar system to a heliocentric one, with an accompanying reduction of the hubris in thinking that the universe was made for us, a decentred ontology of science may reduce our hubris in our thinking about the role of humans as standing in some way apart from the natural world. As the aphorism commonly attributed to Mark Twain would have it: 'It ain't what you don't know that gets you into trouble. It's what you know for sure that just ain't so'. In communicating science to students, it is surely simpler to tell them that scientific knowledge is absolute, than to tell them the numerous conditions under which such knowledge is limited, and how these conditions may vary in ways that are not even in principle foreseeable. Yet, what might be the collective costs of this mass miscommunication, especially when science and technology underwrites so much of the way we live and provides the working metaphors for the manner in which we understand how social institutions are supposed to function?

When we acknowledge that the practice of science is an interaction of human and material agency, we gain the insight that the current best science can (and indeed often does) overlook lines of investigation simply because this interaction is a product of the unique human perspective performing the investigation and the material response. Human experimenters derive their intention in an open-ended interaction with the experiment and not in some cold, calculated, hyper-rational manner. Especially from a perspective interested in the development and growth of scientific knowledge, we should want our students to understand the limitations that emerge from these tentative interactions of time, space, people, and phenomena. If we take up an understanding of the tentativeness of scientific knowledge, there is renewed value in students performing practical investigations, not as a means to convince students of the scientific consensus, but as a means to experience for themselves the subjective and personal qualities involved in the making of a scientific claim.

To be sure, such an approach certainly feels different from the standard science instruction that tends to emphasise our mastery over nature. However, it seems apparent that an approach that emphasises humility has a better chance at attending to the collective condition of stupidity I started this chapter with. If we collectively have a misperception of what science *is* and what might be the limits of its reach, would it really be any wonder that we make mistaken decisions and extend broken metaphors when we look in admiration to the successes of science and attempt to emulate it in other contexts?

How might such an approach to science education look like?

What I have discussed in this chapter mostly concerns practicing scientists working at the boundaries of knowledge, for whom these issues are particularly prone to occur. While it might be assumed that for students working to replicate known theories or to simply learn to use basic laboratory equipment, the opportunities for material agency to interfere with experiments may not be prominent, the reality is that such manifestations can be closer than we typically imagine. Even in something as mundane as mensuration, any attempt to increase accuracy and/ or precision necessarily brings us into close contact with the uncertainty posed by material resistance. What I want to introduce in this section is a particular approach to practical investigation that has the potential to bring students into closer contact with non-human agency. As it currently takes the form of a tournament for highly motivated students, it does not yet enjoy widespread adoption; nonetheless, as I will attempt to show, this activity has the potential to deliver a qualitatively superior educational experience for students.

The International Young Physicists Tournament (IYPT) had its early origins in 1979 in the former Soviet Union as a regional competition. It eventually became international in 1988, but mostly attracted participants in (Eastern) Europe. It held its first competition outside of Europe in 2004 (in Australia), and in the most recent competition in 2019, about 170 students from 34 countries participated. Four main features of the IYPT that distinguish it from other competitions. The IYPT has: (i) a strong focus on empirical investigations, which (ii) participating teams have up to a year to work on; (iii) the challenge format pits three teams together in joint contention over the accuracy of the physics; and most significantly (iv) the problems that teams work on pose a unique challenge. The problem set (of about 17 problems) is released a year in advance of the next tournament, at the conclusion of the current one. Prospective participants, typically screened by national education agencies in collaboration with universities, will then have the year to work on the problems, before being selected at local competitions to represent countries. At tournaments, three teams compete: the first reports their findings, a second seeks to challenge the first on the scientific value of their presentation, and a third reviews the first two teams' performances. Additionally, a set of judges, typically university faculty and school teachers, will subjectively score all three teams on a 10-point scale.

However, what might be the crucial piece of the puzzle for the IYPT is the nature of the challenge problems. These problems are non-trivial, pitched at a level that most undergraduate physics majors would find fairly challenging to respond to, even though most of the IYPT participants tend to be approximately grade 12 or equivalent school students. The problems are characterised primarily by their being practical problems that usually lack a complete analytical solution, being 'solved' by practicing physicists often in numerical simulation. For instance, the first five problems of the set for the 2020 IYPT are:

1. Invent Yourself
   Design an instrument for measuring current using its heating effect. What are the accuracy, precision, and limits of the method?
2. Inconspicuous Bottle
   Put a lit candle behind a bottle. If you blow on the bottle from the opposite side, the candle may go out, as if the bottle was not there at all. Explain the phenomenon.
3. Swinging Sound Tube
   A Sound Tube is a toy, consisting of a corrugated plastic tube that you can spin around to produce sounds. Study the characteristics of the sounds produced by such toys, and how they are affected by the relevant parameters.

4. Singing Ferrite

    Insert a ferrite rod into a coil fed from a signal generator. At some frequencies, the rod begins to produce a sound. Investigate the phenomenon.

5. Sweet Mirage

    Fata Morgana is the name given to a particular form of mirage. A similar effect can be produced by shining a laser through a fluid with a refractive index gradient. Investigate the phenomenon.

As can be seen, these problems are simply stated and are amenable to several possible levels of analysis. At the most basic level, it can be possible for novice participants to attempt simple qualitative descriptions of physical relationships, or in the case of Problem 1, a basic prototype. At an intermediate level, some mathematical characterisation of the problem may be possible, with the kinds of analyses of general trends that are accessible using the methods usually taught in schools. Certainly, at the international competition, such levels are not competitive; at the advanced levels, participants instead need to demonstrate good error analysis, provide significant theoretical explanations for any possible deviations from ideality, and often accompany these findings with sophisticated numerical models. However, what is truly novel about these problems is the degree to which the problems have the potential to teach the students about the complex nature of empirical investigations. Instead of supporting the typical 'rhetoric of conclusions' as practical demonstrations of theory, these problems invite students to find and experience for themselves the ragged boundaries between the well-understood physics of the controlled and limited cases and the complex reality of the physical phenomena as it presents itself to human agents.

For instance, some colleagues and I followed a group of three students as they worked on problem 1. Technically speaking, this was mostly an engineering challenge, in that the main goal of this problem was for them to put together a working apparatus rather than to try to characterise a phenomenon. Nonetheless, all of the problems will require students to go through a similar series of experiences and confront the tenuous processes of creating linkages between phenomena and its representation. For this problem, it would seem that the links were created in the reverse manner, in that the students started with fairly well-known relationships between current and its heating effect in conductors, the cooling behaviour of materials, and even schematics for possible set-ups that have been widely shared on the internet. Then, they had to interpret these representations and translate them into practical apparatus which worked, and show, in relation to calibrated equipment, that their apparatus worked. Potentially, to be competitive, these students had to further work out the sources of error, in order to characterise their apparatus and obtain quantitative measurements of these errors.

Among possible designs, the students we observed chose to construct what was called a hot wire ammeter. Such a device has a conducting metal wire held

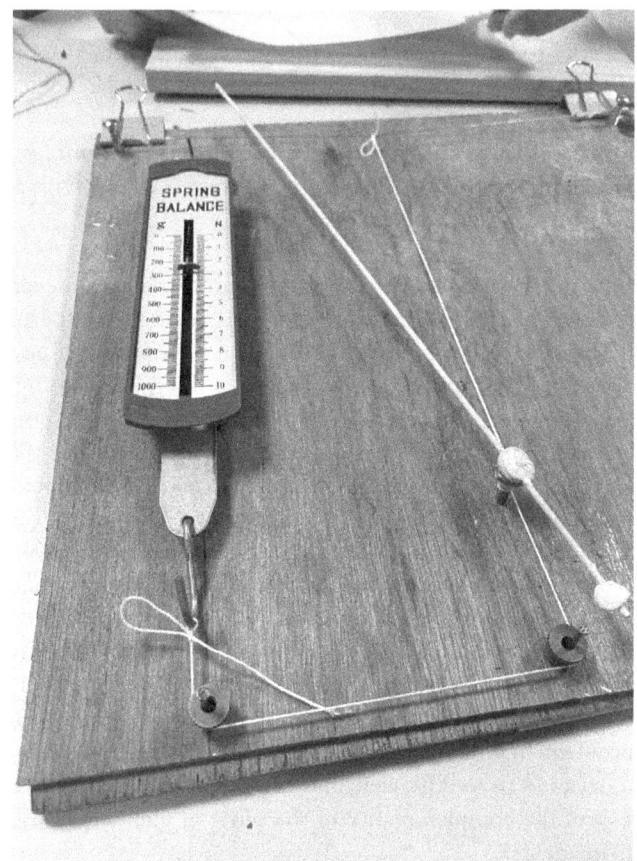

FIGURE 6.1 Final appearance of the experimental set-up created by students

under tension, through which a current (that was being measured) would pass through. A string would pull the midpoint of this wire perpendicular to its main axis, such that this basic arrangement would make a T-shaped layout. By means of winding the string around a capstan like device, the amount of expansion in the conducting wire could then be translated through the deviation in the string, into a rotation of a pointer. See the picture in Figure 6.1.

The wire can be seen near the top of the board, attached to two screws. A string was attached to the wire's midpoint and ran through the improvised capstan to which the wooden skewer was attached as a pointer. The spring balance was there to maintain a repeatable degree of tension in the string.

To be sure, this picture only depicts the layout quite close to the final version that the students had settled on. In order to get to this state, they had numerous prototypes, from rather rudimentary ones that were influenced by so many effects that it almost certainly ensured that no practical value could be derived from it, to intermediate ones which saw promise but were ultimately hampered by bad decisions about the set-up. As could be predicted by Pickering's concept of the mangle of practice, and in a manner which closely mirrored the steps that Glaser took to get his early bubble chambers to work, these students tried something, encountered material resistance in the form of results which did not meet their expectations, made accommodations to these results. All the while, they responded to 'feedback' from the material in ways that gradually minimised extraneous influences, amplifying the desired effect and attenuating the distractions. Eventually, the picture above represents the interactively stabilised apparatus that allowed the students to make claims about how well they were able to harness a particular physical effect for the purposes that they desired.

In this case, the property of the IYPT challenge questions that provided this opportunity for this form of learning was the manner in which there were competing physical effects. For instance, in order for the set-up to provide a readily observable physical effect, it was important for all the heat that was generated to be trapped within the material, such that any tiny amount of heat that was generated could be translated into other form and made 'visible'. On the other hand, if all the heat generated was actually trapped, the temperature of the material would continually increase, and there would be a runaway effect that could not be easily measured. More generally speaking, the IYPT problems all contained some element of complexity or indeterminacy that arose from ill-specified parameters, or physical phenomena that needed to be coaxed into existence. Instead of well-prepared experimental equipment designed to minimise the ways in which non-human agency could affect the outcome of investigations, these problems instead featured the as much variability as might be expected for a physical investigation carried out by a practicing physicist. In this regard, some readers might object that such an approach may not be completely appropriate for a school setting where a majority of students will be of varying motivation levels, not all of which may be interested in untangling the complexity of a realistic investigation.

It might be argued in this particular case that as the concept of the mangle was not formally intended as a learning goal for the students, they may not have acquired it. However, in discussions with them after they completed their presentation at the local qualifying tournament, they nonetheless managed to reflect on their experiences as an insight into the manner in which scientists' claims are often far more limited and circumspect than what might be portrayed in the headlines of popular media. It could be argued here that these IYPT-type challenges at least have the potential for students to acquire more than just the scientific consensus knowledge of how phenomena behaves and that effective facilitation from an instructor can bring students to a different kind of learning not currently extant in school science.

Perhaps, then, the core contention here is the value that might be derived from authentic investigations—as science educators, the temptation to communicate the scientific consensus as a priority can be very high. Yet, while doing so, science educators should contend with what can be lost: opportunities for students to understand how scientific knowledge is actually generated, the explicit and tacit rules that scientific communities use to govern the creation of scientific truth claims and the limits of our knowledge. It is unfortunate that science educators may be mistakenly responding to desires for efficiency. Surely, such an investigation, taking perhaps weeks to finish, and requiring hours of work per session, would not fit into the typical regime of the school day, with its fixed schedules and regular assessments. Yet, I would contend that this is more an indictment of the structure of a public schooling designed for an industrial era that produces proletarians divorced from their knowledge, rather than an education system which is interested in nurturing creators of knowledge and artefacts.

To do this we need a makerspace

In carrying out their investigations, the IYPT challenge does not prescribe specific apparatus or even particular physical designs. Among other competitors to the local tournament was an entry that used the heating effect to cause a bending that was magnified by a laser reflecting off a mirror, for instance. The students we observed had the privilege of using almost anything they could get their hands on from the laboratory storeroom and more if they requested it from their teacher. Still, because of time constraints of waiting for things to be delivered, they made do with things that they essentially scrounged around and found in the store and in the art studio which they visited to look for the building materials such as the plywood board that they used for the backing of their experiment.

In supporting such an activity, it became clear for us that the typical science laboratory was insufficient. Typically designed for short activities and demonstrations, most school science laboratories do not have the space to store in-progress projects. These in-progress projects tend to be particularly fragile, composed of parts that are loosely assembled, and are not suitable for rough handling. To be sure, makerspaces do not necessarily have to be equipped with project storage, but medium- and longer-term projects are usually closer to the norm in many makerspaces. More significantly, in responding to these open-ended IYPT challenges, the possibility exists for students to use multiple approaches to resolve the diverse array of problems that might emerge. Providing one-size-fits-all apparatus may be good for limiting the array of problems that laboratory instructors have to deal with, but perhaps this is more indicative of the level of deskilling that has taken place in conventional school laboratories? Instead of a master experimenter who can understand the multiple ways in which configurations of equipment and materials can go wrong, schools typically have teachers who understand the theory, but who may not be very good with the experiments, and laboratory technicians whose jobs more closely resemble school custodians who know how to deal with the hazards of the laboratory. In making experimental apparatus for these scientific challenges, a wider variety of tools, materials, processing and experimental techniques are often called upon. For instance, many of the competitors make use of computational methods to aid in their investigations, from microcontroller controlled sensor circuits, to computer vision tools that can 'count' rotational rates (for instance) from high-speed video records of objects in motion.

Generally then, the appeal here is for science laboratories to more closely resemble actual practitioner labs in their approach to making experimental artefacts and knowledge claims. Certainly, a counterargument could be made that novices may find such investigations challenging and that a pedagogised version of these experiments may be a better fit for students. While this may be so, it seems to me that such an argument only focusses on the pedagogical aspect of education and misses the curriculum point I am trying to make. In instructing for particular goals such as pedagogical efficacy and efficiency, it is possible to lose sight of what other possible goals that we ought to desire and therefore that we should plan resources to attend to. In this case, I propose that a more genuine approach to investigation, featuring opportunities for instructors to foreground a more accurate depiction of the ontological goals of science, will give societies a greater collective opportunity to deal with its problems that are influenced by science and technology as amplifiers of our intentions. As I mentioned in the opening of this chapter, we may be facing an epidemic of stupidity that is brought about by a drastic misunderstanding of the ways in which scientific knowledge is limited by a nature that is less well behaved than conventional science education would represent. It is this form of enforced ignorance that conventional science education perpetuates. Such an ignorance about the limits of science and its actual creative methods leaves the bulk of school leavers having an overly deterministic idea of what science is capable of, and does societies a disservice.

Makerspaces for school science investigations are thus proposed, not simply as a means to supplement the traditional ways of approaching science instruction, but as a means to reconsider the curriculum goals of science education. Society has changed dramatically since our notion of public schooling has solidified over a century ago; while I will not go as far as to say that a 19th century school teacher would feel right at home in today's classrooms, it is more than likely that this time traveller would not experience dislocation for too long. Much of the change in schooling has been in terms of pedagogical approach; not enough thought, I believe, has been on the goals of schooling, in no small part because curriculum decision-making is often fraught with very overt political contention. It does not help that public schools tend to be conservative organisations, tasked with cultural reproduction as its primary goal. To return to the substantive content of this chapter then, in order for schools to attend to the kinds of open-ended, authentic investigations that practitioners of science would practice, it is vital that activity structures and technological supports more closely resemble what scientists actually use. Certainly, this is not to say that all science education experiences need to strictly replicate scientific practice, but to only provide for students a scaled down, deliberately planned set of experiences that lead students through a metaphorical walk down a walled garden should surely not be enough? Yes, we do want our youngsters not to be poisoned by some herb, eaten by predators, or in the literal case of real makerspaces, cut by knives, but to limit the kinds of learning that is possible seems to either be over-paternalistic or an act of good intentions gone awry.

Conclusion

In this chapter, I have looked at the ontological nature of the practice of science—what science is, what it does, and how truth claims (as social objects) are made in relation to the things of science. Most significantly, the insight that this chapter brings is the way in which material objects can display non-human agency and interact with humans in ways that have not been attended to in most mainstream approaches to science education. In response, I propose that approaches to science education should also give students opportunities to experience the ways in which non-human agency interacts with humans as we set about to discern patterns in nature. This, of course, sets us into obvious conflict with more conservative interpretations of the purpose of schooling as reproducing the established orders of knowing and being. As we become aware that even 'inanimate' objects can react in non-deterministic ways in response to human intentions, we should think again about the process of education and how we wish to bring our students into states of development. It is important for the sake of unforeseeable futures that we preserve students' abilities to think otherwise, and to do so in a manner that nonetheless is respectful of what has come before. The intention here is not to support a pure 'discovery learning' type of instruction where students discover for themselves the theoretical abstractions through a series of activities that resemble how the original investigators practiced. Students need to be told or brought through a series of instructional experiences that researchers have reliably determined to bring about particular conceptions. Yet, there is also a need for students to discover for themselves just how tentative and reliant on a natural world that is not entirely free from capricious behaviour. This is not easy, and it is through experiences in makerspaces that I propose such an approach can be conducted.

Note

1 Reputedly, the Three Gorges dam in China, in collecting 39 trillion kg of water up to 176 m above sea level, slowed the Earth's rotation by 0.06 µs. The effect is similar to a skater who puts their arm out to slow down a spin—raising that much water away from the centre of the Earth causes it to slow down.

References

Alvesson, M., & Spicer, A. (2012). A stupidity-based theory of organizations. *Journal of Management Studies, 49*(7), 1194–1220. Retrieved from https://doi.org/10.1111/j.1467-6486.2012.01072.x.

Arvidsson, A. (2006). *Brands: Meaning and Value in Media Culture.* London: Routledge.

Barad, K. (2003). Posthumanist performativity: Toward an understanding of how matter comes to matter. *Signs: Journal of Women in Culture and Society, 28*(3), 801–831.

Deleuze, G., & Guattari, F. (2004). *A thousand plateaus.* London: Continuum.

Firestein, S. (2012). *Ignorance: How it drives science.* Oxford: Oxford University Press.

Franklin, A. (2008). A choreography of fire: A posthumanist account of Australians and Eucalyptus. In A. Pickering & K. Guzik (Eds.), *The mangle in practice: Science, society, and becoming* (pp. 17–45). Durham, NC: Duke University Press.

Gallagher, S. (2019). They didn't buy the DLC: feature that could've prevented 737 crashes was sold as an option. Retrieved 25 September 2020, from https://arstechnica.com/information-technology/2019/03/boeing-sold-safety-feature-that-could-have-prevented-737-max-crashes-as-an-option/.

Gambetta, D., & Hertog, S. (2016). *Engineers of Jihad: The curious connection between violent extremism and education.* Princeton, NJ: Princeton University Press.

Goodwin, C. (1994). Professional vision. *American Anthropologist, 96*(3), 606–633.

Hacking, I. (1983). *Representing and intervening: Introductory topics in the philosophy of natural science.* New York, NY: Cambridge University Press.

Harding, S. (1991). *Whose science? Whose knowledge?* Ithaca, NY: Cornell University Press.

Ingold, T. (2013). *Making: Anthropology, archaeology, art and architecture.* Abingdon: Routledge.

Irwin, R. (2020). Heidegger and Stiegler on failure and technology. *Educational Philosophy and Theory, 52*(4), 361–375.

Kitroeff, N., Gelles, D., & Nicas, J. (2019, October 2). Boeing 737 Max Safety System was vetoed, engineer says. *The New York Times.* Retrieved 25 September 2020, from https://www.nytimes.com/2019/10/02/business/boeing-737-max-crashes.html.

Kornberger, M. (2010). *Brand Society: How Brands Transform Lifestyles and Management.* Cambridge, MA:Cambridge University Press.

Kuhn, T. S. (1996). *The structure of scientific revolutions* (3rd ed.). Chicago, IL: University of Chicago Press. (Original work published 1962.)

Latour, B. (2005). *Reassembling the social: An introduction to actor-network-theory.* Oxford: Oxford University Press.

Lefkow, C. (2019). Ethiopian Airlines crash: What is the MCAS system on the Boeing 737 Max 8? Retrieved 25 September 2020, from https://phys.org/news/2019-03-ethiopian-airlines-mcas-boeing-max.html.

Longino, H. (1990). *Science as social knowledge.* Princeton, NJ: Princeton University Press.

March, J. G. (2006). Rationality, foolishness, and adaptive intelligence. *Strategic Management Journal, 27*(3), 201–214. Retrieved from https://doi.org/10.1002/smj.515

Pickering, A. (1995). *The mangle of practice: Time, agency and science.* Chicago, IL: University of Chicago Press.

Pickering, A. (2008). New ontologies. In A. Pickering & K. Guzik (Eds.), *The mangle in practice: Science, society, and becoming* (pp. 1–16). Durham, NC: Duke University Press.

Pickering, A., & Guzik, K. (Eds.). (2008). *The mangle in practice: Science, society, and becoming.* Durham, NC: Duke University Press.

Polanyi, M. (2009). *The tacit dimension.* Chicago, IL: University of Chicago Press. (Original work published 1966)

Ross, D. (2020). From 'Dare to Think!' to 'How Dare You!' and back again. *Educational Philosophy and Theory*, *52*(4), 466–474. Retrieved from https://doi.org/10.1080/00131857.2019.1678465.

Ryle, G. (1946). Knowing how and knowing that: The presidential address. *Proceedings of the Aristotelian Society*, *46*(1), 1–16. Retrieved 24 April 2019, from https://doi.org/10.1093/aristotelian/46.1.1.

Schaper, D. (2019, June 26). FAA Finds New Problem with 737 Max Jets, delaying their return to flight. *NPR*. Retrieved 25 September 2020, from https://www.npr.org/2019/06/26/736430419/faa-finds-new-problem-with-737-max-jets-delaying-their-return-to-flight.

Stacey, K. (2020). Boeing hid design flaws in 737 Max jets from pilots and regulators. Retrieved 25 September 2020, from https://arstechnica.com/information-technology/2020/09/boeing-hid-design-flaws-in-max-jets-from-pilots-and-regulators/.

Travis, G. (2019). How the Boeing 737 Max disaster looks to a software developer. Retrieved 25 September 2020, from https://spectrum.ieee.org/aerospace/aviation/how-the-boeing-737-max-disaster-looks-to-a-software-developer.

7
RETHINKING EDUCATION FOR SOCIAL CHANGE

In this concluding chapter, I want to sketch out an agenda for an education (in science, in particular) with a vision that attends to the social issues of our times, as well as it recognises its historical location. Such a project is important when we recognise that education is essentially a wicked problem—whenever we attempt solutions to context bound problems we inevitably sow the seeds for problems to emerge subsequently in response to our solutions. A historically informed view also attends to the problem posed by the hypothetical thought experiment based on an experiment done with some primates. In the thought experiment, a group of primates in an enclosure were given access to a ladder that would bring any one of them to access a sweet treat located at the ceiling. However, experimenters were ready; once any one attempted to climb the ladder, all of them were doused with ice water, causing all of the primates to attack the climber. This then set up a pattern where the primates would enforce the norm: no use of the ladder to get to the treat, even though the experimenters stopped turning on the ice water. Sometime later, one of the primates was swapped out for a new one which had no idea of the rule. Predictably, it would attempt to climb the ladder, but the others would put a stop to it. One by one, the primates that had experienced the ice water would be swapped out with new ones that had not, and the pattern continued where the new ones would be inducted into the cultural pattern that the ladder was not to be used. Eventually, all the primates had no experience of the ice water shower, yet none of them would care to try to use the ladder. The lesson for us humans is clear—how often have we encountered situations where, in order to support a mindless adherence to tradition, we have seen invented justifications? Most disappointingly at the time this is being written, must be the ways in which some fractions of once-powerful groups seek to retain or regain power, often in underhanded ways.

The situation in education is somewhat analogous, in that, by its position as a conservative institution invested in reproducing sociocultural patterns of

DOI: 10.4324/9781351116220-7

behaviour, we run the risk of concentrating on efficiently reproducing sociocultural patterns of thinking, without sufficient though as to whether we ought to be doing so. This, unfortunately, also applies to the situations where we are presented with ostensible changes, as with the burgeoning maker and science, technology, engineering, and mathematics (STEM) movements that have swept through many parts of the world. The appropriate metaphor for much of these changes is that of the cargo cult. Based on the behaviour of Melanesian island cults which fetishised the 'cargo' that recent visitors to the island brought with them on planes and ships, they modified existing religions to adopt cult like rituals praying for the return of cargo. The comparison here is made with schools and other well-meaning institutions who seem gripped by the belief that (almost ritualistically) adopting makerspaces and STEM into the curriculum of schools will somehow magically lead to the arrival of innovative behaviours and subsequent economic rewards. It does not help that makerspaces and a type of education privileging engineering and construction seem to have been made popular by the economically wealthier regions of one of the most successful economies in the world. As with the cargo cults' misunderstanding the intent and purpose of the foreign practices, the question that this chapter seeks to address is the manner in which makerspaces, as sites for STEM instruction, should be run, so that we might avoid cargo cultism. Even more significantly, is the question of intent, purpose, and desire—if we understand why particular outcomes are desired, it becomes possible to use contextually relevant instructional techniques to achieve these intentions. The plan for this chapter is as follows: I will introduce the notion of the cargo cult and inquire if makerspaces and the STEM movement have the potential to become cargo cults. I will then discuss what ideals we should desire and, finally, suggest some principles for which these ideals may be achievable.

Makerspaces as cargo cult

For makerspaces and STEM education, the temptation is to focus on the technological artefacts; with the proliferation of things, it is easy to acknowledge that teachers can feel overwhelmed with communicating the instructions of how these artefacts are supposed to be used. Yet, I want to assert that the more valuable prize for makerspaces and instruction in STEM is a consideration of the appropriate instructional approaches that we should adopt if we desire innovative behaviours as an outcome. In this regard, the sociocultural environment of the makerspace needs close attention, and makerspaces should be considered a form of cultural innovation rather than the physical innovation of rearrangement of furniture and technologies. We have seen in a previous chapter that things have politics: our use and deployment of technologies are not neutral decisions. For instance, if two teachers claim that all students are welcome to participate in makerspace activities, but one keeps essential tools hidden away from student use 'as a precaution', while the other makes them all accessible, it should be clear that these teachers mean different things. In this respect, it is not enough for teachers to make visits to prominent makerspaces and to copy equipment and material lists, without understanding *how* these things are to be deployed.

The situation here is comparable to the cargo cults of Melanesia that explorers from technologically advanced societies have encountered and inadvertently help produce over the last 200 years or so (Worsley, 2009). As with the aphorism that 'any sufficiently developed technology is indistinguishable from magic', the Melanesian peoples failed to understand how the advanced technologies that the visitors worked. In combination with mythical and religious beliefs about apocalyptic end times, the Melanesians adopted strange (by our contemporary understanding) practices such as making wooden facsimiles of aircraft, radar installations, and runways. Ritualistically copying the drills that they observed the visitors performing, the islanders attempted to bring back the 'cargo' by appeals to their gods. From their faulty assumptions about the relationship between the practices and the arrival of material possessions, came the notion of cargo cult science, made popular by physicist Richard Feynman (1974). In his speech, Feynman suggested that there were researchers who were involved in investigations that did not pay sufficient attention to its details, and which therefore were not much different from the cargo cults of Melanesia. Worse yet were schools which, for the sake of promoting replicability of results as a goal in itself without thinking about the rationales for such a move, were only promoting students who could get positive experimental findings. Feynman said then that: 'it is very dangerous to have such a policy in teaching—to teach students only how to get certain results, rather than how to do an experiment with scientific integrity'. And the same criticisms apply even today, especially in time limited school classrooms where it is not uncommon to hear of students 'reverse engineering' an experimental result to create spurious data that better fits the trend line that they believe to be correct.

We read about these episodes with a certain degree of bemusement, perhaps even bordering on condescension, but other researchers have used the concept of the cargo cult to describe the wholesale borrowing of policies, artefacts, or innovations in general (Dexter, 1999; Hattie & Hamilton, 2018; Hughes, 2010). Returning to the main argument of this section, with makerspaces and STEM education ideas circulating the education community, it should not be a surprise at all that not all implementations will take place with sufficient attention to detail and the integrity to consider the educational merit of particular courses of action. Indeed, it would be safe to say that a certain cargo cultism is likely to occur even simply because of the vast quantities of new cargo involved. As teachers, the focus of classroom interactions is conventionally on the communication of particular powerful ideas, especially if some sort of standardised test will be around the corner. It seems natural that teachers would regard makerspaces the same way that they view other forms of technology—as labour saving devices, or in this case, as a means to 'engage' students with fun activities and contextually sensitive artefacts that are meant to be relevant. On a policy and curriculum level, discourse is rife around makerspaces as creative spaces that are supposedly modelled after practices of creative design and innovativeness found in economically successful sectors.

As educators deal with young people whose futures are largely underdetermined, we have an occupational hazard of a fascination with the new (Burbules, 2016). Educational technologies fulfil that desire, promising all manner of improvement in the way we teach and learn. Yet, a historical sweep reveals this not to be the case (Cuban, 2001), and despite contemporary enthusiasm in information processing equipment, similar criticisms may be levelled against an overenthusiastic acceptance of the claims of 'ed tech' (Godhe, Lilja, & Selwyn, 2019; Selwyn, 2016a, 2016b). This is not to say that we should reject technologies, but rather consider that technologies are amplifiers—things that do not, in themselves, cause changes, but merely amplify existing cultural arrangements (Toyama, 2015). A tablet computer could be used to access the internet and lead to great learning, or useless distraction; what matters is what people intend to do with it (Chan, 2014). More generally speaking then, is the insight that things can communicate politics, as in bridges made deliberately low such that buses could not travel along a particular road that led to Long Island in New Jersey. Such architecture limited access to a valued beach only to those who had the 'correct' means of transport (Winner, 1980; Wyatt, 2008). Wyatt (*ibid.*) warns us against adopting technological determinism—the notion that technologies are developed in isolation away from social interests, and its flip side, that when thrust upon the world, technologies can change societies purely by its technical merit. The rise of the globalised internet companies would not be possible without the investment of large monied interests (Zuboff, 2019), and government policies tolerant of, or oblivious to, deep contradictions between stated beliefs and actual implementations (Barbrook & Cameron, 1996).

Even though interpretations of STEM are currently in flux, makerspaces have been commonly associated as sites for STEM education. While there are no fixed prescriptions as to what such a space should constitute, it is common to find some combination of hand and powered tools, electronics, robotics kits, essentially a junior mechatronics engineering workshop (Blikstein, 2013a; Sheridan et al., 2014). That these spaces are not modelled after, say, computing-, bio-, or chemical engineering workshops is something that is potentially problematic. The things, activities, and human organisation of such spaces can have very different effects on the learning outcomes, not only in terms of cognitive content, but also how participants perceive the appropriate goals of activity in such a space. Most famously, contemporary hacker culture (of the positive form such as in hackathon, not associated with criminality) emerged from computing labs and tinkering with model railroads in places such as the Massachusetts Institute of Technology (MIT) (Lindtner, 2014; Wark, 2006). Among hackers, mainframe class computers were resources to be shared, made to do different things, and not merely held in reverence (Isaacson, 2014; Raymond, n.d.) Arguably, the foundational values of openness and communal sharing underlie the economic success of the current internet era, with open source operating systems such as Berkeley Software Distribution (BSD), linux and its derivatives driving a different paradigm in software development. In Singapore, at least one school has made use of a localised version of hacker culture to good result (Tan, 2019); on the other hand, others have suggested that some of these values may not be compatible to pro-social orientations to learning because of its inclination towards individualism (Ames, 2018; Blikstein & Worsley, 2016).

In any case, considering innovativeness as an outcome and design as one method for attaining it (Buchanan et al., 2013; Cross, 2006; Fagerberg, Fosaas, & Sapprasert, 2012; Mabogunje, Sonalkar, & Leifer, 2016), understanding the 'human factor' in innovation is essential. Arguably, the labs, workshops, and technical abilities of local STEM practitioners in many of the 'developed' countries that are considering implementing makerspaces and STEM are as well appointed as any in the most productive cities in the world. Yet, due to different goals perceived valuable, the outcomes have been markedly different. The distributed cognition perspective (Cowley & Vallée-Tourangeau, 2017; Heavey & Simsek, 2017; Hutchins, 1995) allows us to perceive things differently. Instead of considering that individual minds require more knowledge or skills, it is the collective ensemble that needs an 'upgrade'. Continuing the computing metaphor, the 'operating system' for collective thought and action are the cultural beliefs and practices of the organisation. This position is also supported by the componential model of creativity (Amabile, 1983, 2012; Amabile & Pratt, 2016; Conti, Coon, & Amabile, 1996): creative outcomes are due to (i) domain-specific knowledge; (ii) creativity general skills; (iii) motivations; and (iv) sociocultural factors that nurture creativity.

In studies of STEM or science-rich workplaces, a focus on the sociocultural aspects of learning is under-researched. Instead of learning being assessed by the amount of abstract scientific/technical ability, what was found more essential were factors like the freedom to pursue one's professional interests, access to expertise among colleagues, and above all past experience in tinkering (Lottero-Perdue & Brickhouse, 2002). Even in situations where individuals largely worked alone, powerful communities of practice where important knowledge circulated and professional identities developed were possible (Orr, 1998). Among nurses, dental and veterinary technicians, it was found that previous school science was largely irrelevant compared to scientific knowledge and procedural skills acquired on the job (Aikenhead, 2005; Chin, Munby, Hutchinson, Taylor, & Clark, 2003). Also, the degree to which individuals learn on the job is directly dependent on the degree of learning afforded to the organisation at large—they are mutually constitutive. What was surprising too was that learning can occur even though one's paid labour seemed routine provided one acted upon one's curiosity to better one's skills—learning can be pervasive (Lee & Roth, 2005, 2007).

In other words, simply bringing in the things and practices of places which have been known to be innovative may not result in similar outcomes. What is necessary for innovation is for the people who practice in the space to *desire* to creative outcomes and a sociocultural pattern of behaviour that supports and nurtures the risky work that is creativity. The risk is not from such safety critical factors as laboratory accidents, but risk in the sense that every attempt at novelty confronts the possibility of failure and the associated threats to one's ego. Such factors are not completely amenable to forms of characterisation that are often associated with the scientistic impulse such as quantification and records of sequences of behaviour. Because of the diversity with which humans may respond to and contribute to localised cultural norms, it is likely that diverse ways of being creative exist. So far, this notion of the cargo cult signposts a possible pitfall in our quest to change education for a different kind of social outcome. It might be appropriate now to consider what kinds of social outcomes we should desire.

What kind of society is ideal?

Education has always been a future oriented statement of how young people are to be prepared to take over after we die. The choice of this particularly morbid phrasing is deliberate, as at once a visceral reminder of the fundamental unity of the human experience, and a means to focus our thoughts on the purpose of education. Even though dying is a central unifying 'experience' of all human beings, we seldom think about it, instead of believing and acting as if we will be around forever. Posed in these terms, the curriculum question of what knowledge is of most worth becomes more stark: whose ideas do we wish to grant immortality, a relief from the permanence of death, and whose do we wish to forget forever? In these terms then, the binary choice is between education for social reproduction, of keeping these memories alive within the collective, or education for social change, for the purpose of forgetting particular memories. The choice to talk about our mortality here is also associated with the opposite end of the process—babies are born into this world with little innate skills. As with the old myth of Prometheus, we are unlike other species in that we do not have particular talents, but we do have the singular talent of our ability to learn. The curious thing about learning is that while the ones providing the instruction can exert some control over what is to be acquired, the ones doing the acquisition also play a significant role in what they want to learn. In other words, it is possible for learners to refuse. To be sure, such refusal can be rare as instructors usually are in positions of authority and power over learners, but this does not detract from the overall point that, as with Hannah Arendt's notion of natality, every generation gets born anew and brings with it the promise that we might find a way past the quarrels of the old world. Of course, such a position is not always possible, and in fact, as we have seen through historical periods, it can be exceedingly easy for young people to become recruited into older people's ideological struggles, up to and including the kinds of violent radicalisation that can end up in people giving up their lives for the sake of ideas.

If this sounds like an overly melodramatic backdrop to the problem of curriculum and pedagogy for science/STEM education, I would suggest that it is only because we have an unnecessarily narrowed conception of the purposes of science education. Yes, it is important that educators be good at their craft—when we communicate scientific knowledge to students it is important that they acquire it. But to think that education and education research should only concern itself with the technical and technological aspects of this act of communication and learning, and not with the ultimate goals to which we might consciously or inadvertently be leading our students to, is exactly the kind of mistake that an overly specialised, overly rational distribution of labour whose end results may result in irrationality for everyone involved. There has been historical precedent for science educators to consider the wider implications of the knowledge that they are communicating, even if, according to historian of science education John Rudolph (2020), it was only for a brief shining moment towards the end of the 1800s. Referring to it as the 'lost moral purpose' of science education, Rudolph remarks that we have not always considered science education as instrumental for economic development, as it appears to be predominant in recent calls for the increasing emphasis in STEM education. Such calls take the form mostly in terms of how much STEM activities may increase the engagement of students in science and mathematics, ultimately for the preparation of skilled workforces necessary given the numerous technological changes that are current and predicted to occur. In this manner, appeals for STEM and science education have to be seen essentially as threats—learn STEM, or be relegated to economic redundancy in future economies. Given the numerous ways in which contemporary societies are suffering and could potentially suffer from an irrational approach to decision-making that is not grounded in the kind of honest attempts at truth that science represents, Rudolph suggests that what is needed is a reconnection of science with its moral purposes. We should teach science as a means to attend to moral problems; not with the hubris that science can solve all problems, but with the humility of understanding that some problems are the result of science and technology itself, especially when particular intentions are amplified.

Thus, the intention throughout this book has been to try to see the phenomenon of science education, not as an isolated, reductionist cog in the machine of society, in psychological and individualistic terms focussing on 'academic outcomes' as the sole result that matters, but part of the intellectual and political struggle for the bringing about of a different future. To do this, I have had to consider the role of knowledge and truth in social contexts—it is apparent that scientific knowledge, even though socially constructed, still remains our best knowledge to date. We have also had to deal with what might be the fallout of the academic response to the demarcation problem, and the recognition that scientific knowledge may be flawed. To deal with this, I have suggested that we consider the ontology of making and the empirical investigation: as educators, we need to recognise that a narrow minded focus on academic achievement has the tendency to accelerate students 'learning trajectories' towards known rhetoric of conclusions. There is, ultimately, value in the slow and tedious process of fumbling through the 'experimental errors' that will be endemic to experiments conducted with improvised equipment, not only from the embodied sense of grounding the metaphors and concepts in science, in real-life experiences. There is also value in the struggle to obtain truth claims from experiments where investigators have to coax nature in a mutual dance of agency. These aspects of practical investigations ground humans in the nature of truth claims as necessarily partial not in the sense that we make truths up, but that we can only do so in close collaboration with the people and things that make up the scientific practice. If obtaining truth about the universe is a first step in our ability to take action, the second must be the deployment of these knowledge in ways that fulfil particular human intentions. For this, we have the knowledge of design as a transdisciplinary melding of the sciences and the humanities: not only should we know what the current state of nature is, we should also have some idea about how to 'read minds' as in the anthropological and sociological methods of finding out what ails individuals and communities, and wisdom from the humanities to understand what problems are worth solving and what are ethical intentions to have.

In sum then, science and STEM educators and researchers need to stop thinking of their jobs in isolation. 'Academic excellence' is not enough, and at the same time, efforts to directly redress sociopolitical problems such as endemic race/class/gender inequality through the mechanism of public schooling can fail the 'smell test', for me at least, of distinguishing between education and indoctrination. I do not mean to say that social justice oriented approaches to public schooling necessarily falls into the realm of indoctrination. It might certainly be the case that such approaches are well considered and make use of all the legitimate means accessible to public schooling to address sociopolitical problems of its time. But any effort to do so must necessarily involve an act of make believe (Benson & Stangroom, 2006) or engage students in acts of politicisation which may not be ethically defensible (Conroy, 2020). It is one thing to provide access to high status STEM knowledge to students who have been traditionally underserved due to systemic perceptions of who qualifies to do science, but to involve students in quarrels involving unequal distribution of power as a means to engage with the scientific knowledge runs the risk of unnecessarily politicising knowledge. As Conroy (2020) would have it, despite how the world may be unjust, it is not necessarily the place of teachers and even parents to introduce them to the quarrels of the old world. We do not, as a policy, expose our young to sexually transmitted infections or drug addiction, even though they are a routine aspect of adulthood; why then should we expose them to the political struggles of adulthood? For Conroy, as for myself, it appears that childhood brings along with it an obligation from adults to insulate them from particular ideas until such time that they have the resources to effectively attend to them. The prospect for young people in society is that they hold the seeds of promise, that they might be able to find the ways to transcend the quarrels of the old world. This does not mean that we do not tell them about how the world is far from imperfect, but that we might do better if we can focus instead on what ideal states they should aspire to.

How would makerspaces attend to these ideals?

One major problem that requires our collective attention seems to be the excess attention that most of public schooling has on the representation rather than on the phenomena that is being represented. This is likely to be the case as a result of desires for efficiency in communication and a reductionist perspective that presumes that an ability to recite and use these knowledge claims is enough evidence of learning. Yes, learning can be demonstrated by our ability to recall and use these knowledge claims, and it is often more efficient to communicate and assess 'conceptual understanding' en masse. However, working in apprenticeship conditions to acquire the forms of tacit knowledge that accompanies truly understanding the phenomena may be what is necessary. If we want societies to be innovative, if we want our children to be able to surpass the knowledge we have generated, it seems clear that they not only need access to the representations, but they will also need to know the ways in which these representations are cut down versions of the phenomena and the effects of these reductions. To be sure, most formal schooling arrangements already do this, as students start off with concrete experiences in preschool and the early years of schooling, with increasing abstraction into the middle years, before eventually giving undergraduates experiences with knowledge (and artefact) creation, typically in small group settings.

Yet, the main question I would like to pose through this book is whether we should be thinking more about this current model of progression of schooling, especially when we might have lost sight of the educative purposes of public schooling. Especially for Singapore, where this book has been written and, I am sure, many other school systems where the concept of meritocracy in societal distribution of economic opportunities has been attached to schooling outcomes, there are indications that meritocracy has been interpreted by school bureaucracies to mean an enforced scarcity (Harney, 2020). That is to say, if meritocracy demands that there be winners and losers (Labaree, 2010), the pressure will be upon schools to have standardisable comparisons between students that can reliably discriminate between students of 'higher' and 'lower' merit. Schools used to use the mechanism of recitation to determine merit, and while we have expanded the scope of assessment to also include an expanded scope of what constitutes learning that is meaningful, our continued fixation with meritocracy may be holding us back.

As symptom of this problem, consider the typical progression of learning which privileges recall of well-established representations as a logical first stage before students are allowed to attempt creativity. The very real risk exists that:

> There is a point in the training of every scientist in which they make the transition from textbook-oriented learning to discovery, not only of facts, but discovery of questions […] The conversion from an obsessive focus on facts in textbooks to the search for better questions is the point at which

a young scientist becomes an independent actor. [...] unfortunately this is precisely where the nonscientist gets left behind. Having been forced by an outmoded educational system to bear the misery of overweight textbooks, testing absent understanding, and valueless memorisation, they are left with this as their sole experience of science. Worse than the obvious distaste that this produces in so many, it also provides a disastrously distorted view of science as authoritarian, settled, and immutable. Fortunately, science is none of these things. But *who* besides scientists knows this?

(Firestein, 2015, p. 95)

In my experience with schools, students who have been initiated into this sequence of instruction tend to falter when it is their turn to be creative. Many of them will inevitably turn towards their instructors, after years of being told what to do and how to think, and ask them, what next?

Three principles for makerspaces

In what follows, is an idealistic statement of some design principles for the conduct of makerspaces that might attend to the challenges that have been highlighted through this book. To be sure, these principles may not have been directly tested as in experimental settings and may not have been shown to conclusively lead to particular learning gains. However, as should be clear from the tenor of the chapters so far, this has not been the ambition of this book. Nonetheless, these principles have value in that they derive from well-reasoned arguments, and they are not intended to provide direction in terms of exactly how one might get to the end goals. Rather, these principles serve as curriculum goals for makerspaces, for which significant pedagogical expertise is needed to translate these goals meaningfully into contextually bound educational experiences.

Invent, not manufacture

The first principle for makerspaces should be obvious even from the title of the book. We cannot afford schools to be places where particular educational outcomes are mass produced as if along a production line. This recommendation is not a particularly easy one to propose as the industrialisation of schooling has produced significant outcomes for societies, the result of which is that we have become reliant on particular ways of 'doing school' and interpreting what schools are good for. As above, schools as sites of gatekeeping for an ideological version of meritocracy can mean irrational outcomes such as the reproduction of esoteric and useless facts. Even worse, a possible outcome of such a system is an arms-race condition where assessment writers prepare means to trip up students with increasingly obfuscated lines of questioning, while students learn test-taking skills for these synthetic problems which may not have any relation to how these knowledges are used in the 'real world'. To be fair, a pivot to creativity/innovativeness as an outcome for schools can similarly be used as a means for discrimination, and there is every possibility that the subjective assessment of creative outcomes may become yet another means of selecting for particular kinds of individuals, but if this is the case then the problem is more likely to be with the perceptions of the function of schools, rather than with school itself. I suppose I am sanguine about the prospects of systemic change of the function of schools, and so the invention-centred makerspace will perhaps not form the majority of the educational experience of students.

By focussing on invention as a goal for makerspaces, we attend to the problem of preparing individuals for a future yet unseen. As the saying would have it—the only way to predict the future is to invent it—and so our students will need some exposure to become the inventors of tomorrow. Scientific knowledge, as I have surmised, is necessarily a representation that reduces the full spectrum of the experience of the phenomena. Inventing as a means to create new artefacts to assist in the creation of new knowledge stands as one of the few real experiences in a student's life, especially if we compare against the predominant school paradigm of recitation, commonly of abstract representations. For makerspaces in schools, the form of creativity that is desired is not necessarily of the form that will be economically valuable or socially desirable at the largest scales, but merely something that is new for the student. Creating an artefact or a new knowledge claim can get students to confront the social aspect of creation—should something be brought into being? Should certain knowledge claims be made, and if so, what purposes are served by doing so?

Creating and maintaining an invention-centred makerspace can be particularly challenging to teachers who have become experts in industrialised process of schooling. Being a teacher in such a makerspace will require teachers to upskill—there are holistic sets of skills required to facilitate and supervise student activity in makerspaces, especially if teachers need to improvise to help students achieve what they desire. Doing so might require that teachers develop a greater craft like appreciation for their things and the makerspace as auxiliaries to their mind and amplifiers of their instructional intention. Indeed, some of the more successful makerspace instructors are masters of their spaces, knowing where everything is kept and how tools are to be used; in many cases, they can be the only ones who can work particular finicky equipment; or, as living exemplars of the kind of improvisational competence usually demonstrated in makerspaces, be the only ones who know how the jury-rigged devices work. This makes them simultaneously irreplaceable and bound to particular sets of practices and things, and may be perceived as a boon or a bane.

Focussing on invention as a goal for schooling shifts out collective view of what school can be for, and may help us collectively attend to the problems that exist. Certainly, I am not a proponent of a narcissistic form of 'creativity' that celebrates any form of deviation from established norms. There needs to be some meaningful and intentional change, and there needs to be some form of attention to what has already come before. As with Hannah Arendt (1961), whose notion of natality gives us a means to think past the current quarrels of the old world of political contention, we have to provide schooling opportunities for our young that preserve their abilities to think otherwise, so that they can act otherwise. To do so, we might choose to be careful about what we might be doing with our students (I hope it is appreciated that education is not something that is done *to* someone). The temptation, as with the approaches to science education that use contextually grounded events as a means for instruction Socio-Scientific Issues (SSI), Socially Acute Questions (SAQ), and Science Technology Societies and Environment (STSE), would be to 'expose' students from an early age to the kind of political activism that they would be expected to face later in life. Yet, caution must be advised: just as young seedlings might need a particular nurturing environment to successfully grow to such a stage that they can independently tolerate the diverse environmental conditions that might be occurring, it may not necessarily be a good thing for us to expose our young to the corrosive political contentions of the world of adults (Conroy, 2020). If we seek social change and improvement, it seems to me apparent that we should want those that will be taking over the world to have the opportunity to practice doing so from an early age. In this sense, I view science education in makerspaces as an opportunity for students to practice making a world for themselves that does not repeat the mistakes of the old world. Yes, there is every chance that they fail, but what chance will they have if we will only allow a reproductive approach to schooling?

People, not things

It should be fairly obvious that the attention in makerspaces should be on the people using the space. Yet, it can be possible that instructors spend so much time teaching students how to use the things that they don't do enough to educate students on the social aspects of how the things are supposed to be used. This can especially be so because of the numerous things that would be found in the makerspace and the temptation of instructors to retain some sort of authority in the classroom by 'owning' private knowledge about these things which they can dole out in exchange for compliance to desired behaviours. In advocating this goal for makerspaces, the notion here is to break through individualistic and individualising practices that can be typical of schooling. Yes, we have our thoughts in private, but it is within the community of those whose company we keep that we best develop our minds. Constructionism has, for the last several decades, asserted as much (Ackermann, 2001; Kafai & Resnick, 1996; Papert, 1993), that humans learn best when we make artefacts and re-presentations of our ideas to one another. To date, the constructionism movement appears to have been strong adopters of the makerspaces as sites for learning, and many researchers have made use of constructionism as a means to understand the learning that goes on in makerspaces (e.g., Blikstein, 2013b; Martin, 2015; Sheridan et al., 2014; Vossoughi & Bevan, 2014). While constructionism as learning theory is probably a great tool to think about the kinds of learning processes, and therefore the kind of pedagogical techniques that one should privilege in makerspaces for particular kinds of outcomes (the 'how' question), it may not be well suited to thinking about why certain goals should be pursued, or what the range of goals should be, in the first place. Unfortunately, criticism has also been raised about the manner in which embedded in constructionism are also assumptions about the ideal kinds of learner and social environment (Ames, 2018; Blikstein & Worsley, 2016). Constructionism has been traced to early notions of 'hacker culture' and from there a link to a certain degree of excessive individualism and anti-authoritarianism that can shade into an excessive libertarianism.

As I have written earlier, we need to do a better job of attending to public perceptions of technological determinism—that pernicious notion that technologies are developed, apart from societal influence, and emerge fully formed into the world and change society, often for the better. Science education should stop with the excessive specialisation and presume that questions of value and use are 'not science' and do not have a place in the science classroom. This is especially so given how widespread science and technological innovations have become part of our everyday existence, and how entropic asymmetry can play in the hands of malfeasant individuals. With entropic asymmetry, we recognise that with complicated systems, there are more ways for the system to be broken, then there are ways for the system to be working properly (Reid, 2019). In addition to entropic asymmetry, is the terrifying notion that many of the technologies that can be subverted for evil are already within easy grasp of individuals. Even for myself, the parts and instructions to create a so-called 'TV-b-gone' can be obtained for very little money. This device works as a universal remote that sends out infrared signals for the 'turn-off' code for about a dozen major brands of television manufacturers. While this particular device causes little more than inconvenience for people affected, it should be emphasised that for many other technologies and systems, we rely on very little else than a form of security through obscurity, believing that systems are safe because few people bother to interfere with their proper function.

This is not necessarily a call for additional regulation, although that may be called for. Regulators seem to always be on the back foot compared to innovators, and so a scientific/technological hippocratic oath of sorts needs to be part of the education of scientists and technologists. What needs to be part of this curriculum should be attempts to correct the dominant approach to risk distribution of contemporary science and technology projects—as Reid's (2019) interviewee surmises, we used to live in a time where STEM projects had a high public upside and a very private downside. Consider the early developers of nuclear weapons: there was initial reluctance to detonate a test weapon on the safety fears that the entire atmosphere may be consumed by the fireball; additionally, development of the nuclear device involved risks which were borne by the developers, for a reward that was public and did not confer much benefit to the individuals concerned. Today, we have providers of communication infrastructure which embed themselves in the midst, extracting vast amounts of personal information from the public, and transforming all of these information into very private profit. Certainly, there has been an exchange of service in return for the profit, but any risks to this transaction are completely borne by the public, with very little risk for the corporation. Even though these social media firms have selectively amplified particular messages to increase platform 'engagement' and have utilised scientific findings from psychology and other fields to improve user addiction to their platforms, they have so often hidden behind claims of the opacity of their seemingly neutral 'algorithms', all while misinformation, rampant tribalism, and the fracturing of society have taken place, to their profit. We need to nurture scientists and technologists of tomorrow with a pro-social moral compass, and this will not happen for as long as continue to pretend that ethico-moral issues are 'not science'.

Science education in makerspaces can develop the kinds of social orientations that not only help students learn, but also make the abuse of science and technology less likely. This can come about through a more proactive monitoring of student dispositions and an increased attention to the kinds of intentions that are expressed through the artefacts that they intend to make. Again, it is not enough to teach 'the science' and assume that the ethical aspects will be attended to 'in another class'—this has to be a holistic experience for our students. As for teachers, this expansion of duties and responsibilities may seem onerous, and many will complain that these aspects are not part of science teachers' jobs. Yet, for me, this is more an indictment of the deeply entrenched nature of proletarianised work in schools, which needs our collective attention to solving. Yes, the industrialisation of education, as with the industrialisation of so many other sectors of production, has expanded access to luxuries formerly reserved for a small coterie of the elite. However, we might want to be more careful to consider if we will be willing to pay the costs associated with such an expansion. If the industrialisation of schooling dehumanises its participants, if it creates a compartmentalisation that results in terrible consequences such as the inability of collectives of individuals to see the 'big picture' because we are too busy arguing past one another on the details of our piece of the puzzle; if it does any of these things, perhaps we might want to think of a better way forward?

Just in time, not just in case

As part of the industrialism that has commandeered public schooling, the decision-making authority as to the forms of knowledge that are to be delivered by the machine of public schooling has been decided by a group of specialists. Quite often, these specialists have been given a political mandate, as it should be fairly obvious to most that schooling of any sort is a deeply political process, a statement that someone else's children should be inculcated one's values. The contention can be obvious in the sciences in cases such as biological evolution, but for the physical sciences (and other sciences, even biological sciences more generally) it is commonly perceived to be unproblematic. We tend to believe that a particular selection, sequencing, and pacing of the curriculum is ideal in order to arrive at particular outcomes at particular stages of a child's development and would wonder what happened to someone's schooling if they did not know certain 'basic facts' at certain ages.

But should this really be the case? It seems to me to be a waste of human abilities to insist that broad swathes of children know a large collection of somewhat arbitrarily selected pieces of knowledge so that perhaps a small fraction of these children will someday be prepared with the intellectual tools to cope with a challenge that may or may not arrive in the future. Certainly, we may be able to predict with some accuracy, based on the trends that currently exist, what scientists and technologists of the near future will need. But, such a strategy surely inadvertently places limits on the kinds of problems that can be understood and the kinds of solutions that are considered plausible. If we were to take the example of Roger Bannister running the four-minute mile: it was widely considered impossible for humans to break the four-minute barrier, until Bannister did it, and in quick succession other runners followed. What seemed to have changed is not the introduction of particular technologies such as new shoes or new training regimens, but the knowledge that it was in fact possible. For scientific and technological innovativeness, a similar mechanism must also exist—that once people start to see that a different way is possible, others will follow.

To educate for the possibility of the outgrowth of new ideas, I believe we need students who at least have had the experience of pursuing a particular programme of research in reaction to something that interests them, rather than always having courses of action prescribed to them. Certainly, there is no intention here to advocate a kind of curricula anarchy, of suggesting that students necessarily will now do better than what is needed by the teachers. We will still need teachers, but not as assembly line technicians attaching pieces of knowledge according to the foreman's instruction. Yet again, I am advocating a craft-based interpretation of the work of teaching, responding with discernment and judgment to the student as a person, with legitimate interests and concerns that can express their preference for how they would prefer to be educated.

Such an approach should be obvious if we were to take up the first two principles—if we are to be true to the development of an inventive mindset, how can it be possible to pre-specify for our students the kind of knowledge they will need to invent what they want? Yes, teachers will want ways to reduce the amount of surprise that they will be exposed to in responding to students' ideas, but that cannot be at the expense of crimping students' creativity. As adults, teachers possess the maturity and experience gathered from a wide ranging exposure to different disciplines. Against the specialisation model of teachers as narrow experts in particular disciplines only, makerspaces will need teachers to serve as concierge, helping travellers to enjoy their vacations in their destinations by offering useful local knowledge about where the best places are, how to get to where they want to go, and generally being helpful to their guests.

Conclusion

The institution of mass public schooling has been a product of its times, and it is actually quite amazing that we have managed to modify it, akin to a contemporary Ship of Theseus, to accommodate the diverse challenges that we have thrown at it over the years. Yet, for the reasons that I have reviewed throughout the book, it should be quite clear that it may be time to buy a new boat. It need not be that we should abandon this ship, for the new boat is still new, and its design is not well tested. Moreover, its carrying capacity is rather low, as not enough experienced sailors know how to sail in this boat yet. For the sake of a future that will not repeat the mistakes of the past, it seems to me wise that we not only try launching a few small boats, but also, indeed, fleets of nimble seacraft that can explore and quickly return to tell the collective where the shoals are and where we might find new unexplored lands. Certainly, this will involve risks, but that is precisely the nature of the human experience—as Theodor Adorno surmised so long ago, but we seem to have forgotten in the intervening years: we humans believe ourselves free from fear when there is no longer anything unknown. This is not to be the case. There will always unknowable futures, which we will attempt to fix either by actively modifying reality to fit our theories of the universe, or by asserting truth by fiat. By embracing risk, by teaching our students how to do so, and by developing better attitudes to risk, hopefully we might also get to see the rewards of taking these risks.

References

Ackermann, E. (2001). Piaget's constructivism, Papert's constructionism: What's the difference. *Future of Learning Group Publication, 5*(3), 438.

Aikenhead, G. S. (2005). Science-based occupations and the science curriculum: Concepts of evidence. *Science Education, 89*(2), 242–275. Retrieved from https://onlinelibrary.wiley.com/doi/abs/10.1002/sce.20046.

Amabile, T. M. (1983). The social psychology of creativity: A componential conceptualization. *Journal of Personality and Social Psychology, 45*(2), 357.

Amabile, T. M. (2012). *Componential theory of creativity.* Retrieved from Harvard Business School: https://www.hbs.edu/faculty/Publication%20Files/12-096.pdf.

Amabile, T. M., & Pratt, M. G. (2016). The dynamic componential model of creativity and innovation in organizations: Making progress, making meaning. *Research in Organizational Behavior, 36*, 157–183. Retrieved from https://doi.org/10.1016/j.riob.2016.10.001.

Ames, M. G. (2018). Hackers, computers, and cooperation: A critical history of logo and constructionist learning. *Proceedings of the ACM on Human–Computer Interaction, 2*(CSCW), 18:1–18:19. Retrieved from https://doi.org/10.1145/3274287.

Arendt, H. (1961). *Between past and future: Six exercises in political thought.* New York, NY: The Viking Press.

Barbrook, R., & Cameron, A. (1996). The Californian ideology. *Science as Culture, 6*(1), 44–72. Retrieved from https://doi.org/10.1080/09505439609526455.

Benson, O., & Stangroom, J. (2006). *Why truth matters.* London, UK: Continuum.

Blikstein, P. (2013a). Digital fabrication and 'making' in education: The democratization of invention. In J. Walter-Herrmann & C. Büching (Eds.), *FabLabs: Of machines, makers and inventors* (Vol. 4, pp. 1–21). Bielefeld: Transcript Publishers.

Blikstein, P. (2013b). Digital fabrication and 'making' in education: The democratization of invention. In *FabLabs: Of machines, makers, and inventors*. Bielefeld: Transcript Publishers.

Blikstein, P., & Worsley, M. (2016). Children are not hackers: Building a culture of powerful ideas, deep learning, and equity in the maker movement. In K. Peppler, E. Halverson, & Y. Kafai (Eds.), *Makeology: Makerspaces as Learning Environments* (pp. 64–80). New York, NY: Routledge.

Buchanan, R., Cross, N., Durling, D., Nelson, H., Owen, C., Valtonen, A.,, ..., & Visscher-Voerman, I. (2013). Design. *Educational Technology, 53*(5), 25–42. Retrieved 8 August 2019, from http://www.bookstoread.com/etp.

Burbules, N. C. (2016). Technology, education, and the fetishization of the 'New'. In P. Smeyers & M. Depaepe (Eds.), *Educational research: Discourses of change and changes of discourse* (pp. 9–16). Cham: Springer. Retrieved from https://doi.org/10.1007/978-3-319-30456-4_2.

Chan, A. S. (2014). Beyond technological fundamentalism: Peruvian Hack Labs & inter-technological education. *Journal of Peer Production*, 5. Retrieved from http://peerproduction.net/issues/issue-5-shared-machine-shops/peer-reviewed-articles/beyond-technological-fundamentalism-peruvian-hack-labs-and-inter-technological-education/.

Chin, P., Munby, H., Hutchinson, N., Taylor, J., & Clark, F. (2003). Where's the science?: Understanding the form and function of workplace science. In *Reconsidering science learning* (pp. 130–131). Abingdon: Routledge. Retrieved from https://www.taylorfrancis.com/books/e/9780203464021/chapters/10.4324/9780203464021-21.

Conroy, J. (2020). Caught in the middle: Arendt, childhood and responsibility. *Journal of Philosophy of Education, 54*(1), 23–42. Retrieved from https://doi.org/10.1111/1467-9752.12367.

Conti, R., Coon, H., & Amabile, T. M. (1996). Evidence to support the componential model of creativity: Secondary analyses of three studies. *Creativity Research Journal, 9*(4), 385–389. Retrieved from https://doi.org/10.1207/s15326934crj0904_9.

Cowley, S. J., & Vallée-Tourangeau, F. (2017). Thinking, Values and Meaning in Changing Cognitive Ecologies. In *Cognition beyond the brain: Computation, interactivity and human artifice*. Cham: Springer. Retrieved from https://doi.org/10.1007/978-3-319-49115-8.

Cross, N. (2006). *Designerly ways of knowing*. London: Springer-Verlag London.

Cuban, L. (2001). *Oversold and underused: Computers in the classroom*. Cambridge, MA: Harvard University Press.

Dexter, S. (1999). Collective representations and educational technology as school reform: Or, how not to produce a cargo cult. *Educational Technology & Society, 2*(4), 53–61.

Fagerberg, J., Fosaas, M., & Sapprasert, K. (2012). Innovation: Exploring the knowledge base. *Research Policy, 41*(7), 1132–1153. Retrieved from https://doi.org/10.1016/j.respol.2012.03.008.

Feynman, R. (1974). Cargo cult science. Retrieved 12 February 2020, from http://calteches.library.caltech.edu/51/2/CargoCult.htm.

Firestein, S. (2015). Sharing the resources of ignorance. In M. Gross & L. McGoey (Eds.), *Routledge international handbook of ignorance studies* (pp. 92–96). Abingdon: Routledge.

Godhe, A.-L., Lilja, P., & Selwyn, N. (2019). Making sense of making: Critical issues in the integration of maker education into schools. *Technology, Pedagogy and Education*, 1–12. Retrieved from https://doi.org/10.1080/1475939X.2019.1610040.

Harney, S. (2020). Meritocracy in Singapore. *Educational Philosophy and Theory, 52*(11), 1139–1148. Retrieved from https://doi.org/10.1080/00131857.2020.1753034.

Hattie, J., & Hamilton, A. (2018). *Education cargo cults must die.* Thousand Oaks, CA: Corwin.

Heavey, C., & Simsek, Z. (2017). Distributed cognition in top management teams and organizational ambidexterity: The influence of transactive memory systems. *Journal of Management, 43*(3), 919–945. Retrieved from https://doi.org/10.1177/0149206314545652.

Hughes, A. (2010). Innovation policy as cargo cult: Myth and reality in knowledge-led productivity growth. In A. López-Claros (Ed.), *The innovation for development report 2009–2010: Strengthening innovation for the prosperity of nations* (pp. 101–117). Basingstoke: Palgrave Macmillan.

Hutchins, E. (1995). *Cognition in the wild.* Cambridge, MA: MIT press.

Isaacson, W. (2014). *The innovators: How a group of inventors, hackers, geniuses and geeks created the digital revolution.* New York, NY: Simon and Schuster.

Kafai, Y. B., & Resnick, M. (1996). *Constructionism in practice: Designing, thinking, and learning in a digital world.* Mahwah, NJ.: Lawrence Erlbaum Associates.

Labaree, D. F. (2010). *Someone has to fail: The zero-sum game of public schooling.* Cambridge, MA: Harvard University Press.

Lee, Y.-J., & Roth, W.-M. (2005). The (unlikely) trajectory of learning in a salmon hatchery. *Journal of Workplace Learning, 17*(4), 243–254. Retrieved from https://doi.org/10.1108/13665620510597194.

Lee, Y.-J., & Roth, W.-M. (2007). The individual| collective dialectic in the learning organization. *The Learning Organization, 14*(2), 92–107. Retrieved from https://www.emeraldinsight.com/doi/abs/10.1108/09696470710726970.

Lindtner, S. (2014). Hackerspaces and the internet of things in China: How makers are reinventing industrial production, innovation, and the self. *China Information, 28*(2), 145–167. Retrieved from https://doi.org/10.1177/0920203x14529881.

Lottero-Perdue, P. S., & Brickhouse, N. W. (2002). Learning on the job: The acquisition of scientific competence. *Science Education, 86*(6), 756–782. Retrieved from https://doi.org/10.1002/sce.10034.

Mabogunje, A., Sonalkar, N., & Leifer, L. (2016). Design thinking: A new foundational science for engineering. *The International Journal of Engineering Education, 32*(3), 1540–1556.

Martin, L. (2015). The promise of the maker movement for education. *Journal of Pre-College Engineering Education, 5*(1), 30–39. Retrieved from https://doi.org/10.7771/2157-9288.1099.

Orr, J. E. (1998). Images of work. *Science, Technology & Human Values, 23*(4), 439–455. Retrieved from https://doi.org/10.1177/016224399802300405.

Papert, S. (1993). *Mindstorms: Children, computers, and powerful ideas* (2nd ed.). New York, NY: Perseus Books.

Raymond, E. S. (n.d.). A brief history of Hackerdom. Retrieved 4 January 2020, from https://immagic.com/eLibrary/ARCHIVES/GENERAL/AUTHOR_P/R000825P.pdf.

Reid, R. (2019). Ars on your lunch break: Let's talk about the extinction of humanity. Retrieved 3 December 2019, from https://arstechnica.com/ars-podcast/2019/06/ars-on-your-lunch-break-lets-talk-about-the-extinction-of-humanity/.

Rudolph, J. L. (2020). The lost moral purpose of science education. *Science Education, 8*, 741. Retrieved from https://doi.org/10.1002/sce.21590.

Selwyn, N. (2016a). Minding our language: Why education and technology is full of bullshit ... and what might be done about it. *Learning, Media and Technology, 41*(3), 437–443. Retrieved from https://doi.org/10.1080/17439884.2015.1012523.

Selwyn, N. (2016b). The Dystopian futures. In N. Rushby & D. W. Surry (Eds.), *The Wiley handbook of learning technology* (pp. 542–556). Sussex: John Wiley & Sons. Retrieved from https://doi.org/10.1002/9781118736494.ch28.

Sheridan, K., Halverson, E. R., Litts, B., Brahms, L., Jacobs-Priebe, L., & Owens, T. (2014). Learning in the making: A comparative case study of three makerspaces. *Harvard Educational Review, 84*(4), 505–531.

Tan, M. (2019). When makerspaces meet school: Negotiating tensions between instruction and construction. *Journal of Science Education and Technology, 28*(2), 75–89. Retrieved from https://doi.org/10.1007/s10956-018-9749-x.

Toyama, K. (2015). *Geek Heresy: Rescuing social change from the cult of technology.* New York, NY: PublicAffairs.

Vossoughi, S., & Bevan, B. (2014). *Making and tinkering: A review of the literature.* Retrieved from Washington, DC: San Francisco Exploratorium.

Wark, M. (2006). Hackers. *Theory, Culture & Society, 23*(2–3), 320–322. Retrieved from https://doi.org/10.1177/026327640602300242.

Winner, L. (1980). Do artifacts have politics? *Daedalus, 109*(1), 121–136.

Worsley, P. M. (2009, May 1). 50 Years Ago: Cargo Cults of Melanesia. *Scientific American.* Retrieved 11 February 2020, from https://www.scientificamerican.com/article/1959-cargo-cults-melanesia/.

Wyatt, S. (2008). Technological determinism is dead; Long live technological determinism. In E. J. Hackett, O. Amsterdamska, M. Lynch, & J. Wajcman (Eds.), *The handbook of science and technology studies* (pp. 165–180). Cambridge, MA: MIT Press.

Zuboff, S. (2019). *The age of surveillance capitalism: The fight for a human future at the new frontier of power.* New York, NY: Public Affairs.

INDEX